Polish Science and Technology Studies in the New Millennium

STUDIES ON CULTURE, TECHNOLOGY AND EDUCATION

Edited by Krzysztof Abriszewski

Advisory Committee
Rev. Prof. Jerzy Machnacz (Pontifical Faculty of Theology Wroclaw, Poland)
Prof. Jan Misiewicz (Wroclaw University of Technology, Poland)
Dr. Annette Deschner (University of Education at Karlsruhe, Germany)
Prof. Alicja Kalus (University of Opole, Poland)
Dr. Mojca Kompara (University of Primorska, Slovenia)
Prof. Henryk Kiereś (Catholic University of Lublin, Poland)

VOLUME 8

Krzysztof Abriszewski / Aleksandra Derra /
Andrzej W. Nowak (eds.)

Polish Science and Technology Studies in the New Millennium

Bibliographic Information published by the Deutsche Nationalbibliothek
The Deutsche Nationalbibliothek lists this publication in the Deutsche Nationalbibliografie; detailed bibliographic data is available in the internet at http://dnb.d-nb.de.

Library of Congress Cataloging-in-Publication Data
A CIP catalog record for this book has been applied for at the Library of Congress.

This publication was financially supported by Adam Mickiewicz University in Poznań and by Nicolaus Copernicus University in Toruń.

ISSN 2196-5129
ISBN 978-3-631-84388-8 (Print)
E-ISBN 978-3-631-88267-2 (E-PDF)
E-ISBN 978-3-631-88268-9 (EPUB)
DOI 10.3726/b20008

© Peter Lang GmbH
Internationaler Verlag der Wissenschaften
Berlin 2022
All rights reserved.

Peter Lang – Berlin · Lausanne · Bruxelles · New York · Oxford · Warszawa · Wien

All parts of this publication are protected by copyright. Any utilisation outside the strict limits of the copyright law, without the permission of the publisher, is forbidden and liable to prosecution. This applies in particular to reproductions, translations, microfilming, and storage and processing in electronic retrieval systems.

This publication has been peer reviewed.

www.peterlang.com

Contents

Preface 7

Andrzej W. Nowak
The Disappearance of the Second World and "Suitcase Science" 11

Krzysztof Abriszewski
What Does Theory Do in the Humanities? 21

Aleksandra Derra
Women Not to Be Seen: The Medical Versus Feminist Thought Style
on Anorexia Nervosa 47

Ewa Bińczyk
The Disinformation Rhetoric of the Twenty-First Century 69

Michał Wróblewski, Wojciech Goszczyński
Polish Smog: Metrological Controversies and Conflicting Ontologies 99

Maria Lompe
Learning to Breath: Issue Mapping on the Smog Controversy in Poland
Using the Web 2.0 125

Preface

The history of science studies in Poland goes back to 1910–1939. This early stage brought at least several great scholars including Florian Znaniecki, philosopher and sociologist, president of American Sociological Association (1953–1954); sociologists and philosophers Maria Ossowska and Stanisław Ossowski; or Ludwik Fleck, microbiologist and philosopher, whose *Entstehung und Entwicklung einer wissenschaftlichen Tatsache. Einführung in die Lehre vom Denkstil und Denkkollektiv* (Genesis and Development of a Scientific Fact) from 1935 influenced Thomas Kuhn's *Structure of Scientific Revolution* and offered an early ethnography of laboratory work, forty years before a movement with such a label emerged in science studies, in Californian fieldworks of Karin Knorr-Cetina, Bruno Latour and Michael Lynch. Michał Kokowski, distinguished historian and STS scholar from Polish Academy of Science, tells detailed history of science studies in Poland, and notes a "destruction stage" in 1950s after the first stage of flourishing. The next development stage of science studies in Poland dates from 1960s to 1980s and ends roughly with the fall of the Eastern Bloc.[1] Political structures undergo deep transformation together with structures of knowledge, and that includes both scientific institutions and professional knowledge. One can but mourn when the disappearance of science studies departments and professional discourse turned out to be just a collateral damage – one among many in the academia – of structural adjustment of the institutions of the former Eastern Bloc to the capitalist world-system. Thus, the twentieth century history of science studies in Poland ends with another destruction stage in the 1990s. Traditionally, the Polish term "naukoznawstwo" used to be translated not as "science studies" but rather as "science of science." Even one of the Polish Academy of Science committees is still translated as "Science of Science Committee."

By stressing this, we want to underline the symbolic moment of rupture. Most authors in this volume started their university studies in 1990s, during the "destruction stage" of Polish science studies/science of science. Consequently, their knowledge came from singular texts by Fleck, the Ossowscy, or Znaniecki,

---

1   Kokowski, Michał. "The Science of Science (naukoznawstwo) in Poland: the changing theoretical perspectives and political contexts – a historical sketch from the 1910s to 1993," *Organon*, 47/2015, pp. 147–237.

and not from any systematic university course presenting the whole discipline (science of science or science studies). Therefore, when studying philosophy or sociology with an interest in studying science, they turned to western Science and Technology Studies (STS), which were gaining their momentum after great successes of the Edinburgh School's strong programme, the ethnography of laboratory, the empirical programme of relativism, and the social construction of technology. As it happened, having abandoned their philosophical training, they became particularly interested in the Actor–Network Theory (ANT), which seemed to offer good conceptual tools to combine philosophical questions with sociological theory, particularly with the sociology of knowledge. At the same time, ANT avoided weaknesses of popular individualistic epistemology insensitive to social and cultural dimensions of knowledge. Back then, ANT seemed also like a good way out of postmodern discussions that were uninspiring and, in fact, half dead by the dawn of the new millennium.

The first two decades of the 2000s witnessed growing interest in STS and, particularly, ANT among Polish scholars. This entailed a series of original books, papers, and translations, including five translations of books authored or co-authored by Bruno Latour: *Politics of Nature* (Polish translation: 2009), *Reassembling the Social* (Polish translation: 2010), *We Have Never Been Modern* (Polish translation: 2011), *Pandora's Hope* (Polish translation: 2013), and *Laboratory Life* (Polish translation: 2020). Moreover, Latour became a major inspiration for numerous scholars in Poland in the humanities and social sciences who were interested in posthumanism or developing their own position within the "material turn." As a result of gaining recognition, the local Polish STS community in Toruń hosted the 2014 Congress of the European Association for the Study of Science and Technology.

Research units, institutes, undergraduate, graduate and doctoral programs devoted to STS research have been established at many universities; each of them, depending on the circumstances and interests of researchers and researchers, has its own specificity. Today, their sheer number and diversity indicate how mature STS has become a research field. An example is the European Inter-University Association on Society, Science and Technology (ESST). It is an association of universities in Europe that offer a Master's degree program in science and technology studies and allow for research exchange. The Faculty of Humanities of the Nicolaus Copernicus University in Toruń joined it in 2017, offering the specialization "The Theory and Practice of Risk Society" (http://esst.eu/specializations/).

Thanks to the efforts of researchers at the Nicolaus Copernicus University in Toruń, the first Polish-language reader in the field of STS was created in Poland, entitled *Reader in Science and Technology Studies*, edited by Ewa Bińczyk and Aleksandra Derra (published by scientific publishing house of the Nicolaus Copernicus University in Toruń in 2014). They include translations of texts by Bruno Latour, Donna Haraway, Karen Knorr-Cetina, John Law, Andrew Pickering, Michel Callon, Harry Collins, Steven Shapin, Ruth Hubbard and Ian Hacking along with an appropriate introduction and commentary.

Furthermore, the first two decades of the 2000s brought new generations of young scholars interested in the study of science. They were able to use experiences of the "1990s generation," now their older colleagues and sometime scientific supervisors. While the older generation is more devoted to theory, the younger researchers often turn to empirical research. The number of STS scholars in Poland is growing, embracing this philosophy and methodology in the humanities, social sciences, political science, cultural studies, feminism, along with the study of Anthropocene and climate change, smog problems, contemporary medicine, medical controversies, hacker cultures, and others. Current stage of STS in Poland – the one strongly inspired by ANT – is mainly linked to Nicolaus Copernicus University in Toruń and Adam Mickiewicz University in Poznań. This volume is a small representation of the growing community of Polish STS scholars. We hope that it would be just the first in a series of such presentations that go beyond one country.

Krzysztof Abriszewski, Aleksandra Derra and Andrzej W. Nowak

Bibliography

Kokowski, Michał. "The Science of Science (naukoznawstwo) in Poland: The changing theoretical perspectives and political contexts – a historical sketch from the 1910s to 1993," *Organon*, 47/2015, pp. 147–237.

Andrzej W. Nowak
Translated by Nelly Strehlau

The Disappearance of the Second World and "Suitcase Science"

Abstract: The chapter captures two intertwined problems. The first is the consequence of the transition from describing the world through the prism of the categories of First, Second, and Third World and replacing it with the North-South opposition. The second problem is the question of doing science, and especially science and technology studies, from the semi-peripheral perspective. This is called "suitcase science." The chapter concludes by pointing out the consequences of geopolitical changes for the location of local technoscientific practice in the semi-peripheral countries (Central and Eastern Europe).

Keywords: Semi-periphery, Second World, world-system, science technology studies

In what ways practicing post/socialist Science and Technology Studies (STS) in Eastern Europe might be different from that in the "West," considering the common problems facing academics in Eastern Europe, and how could uneven hegemonic relations be countered? In reply to this question, I would like to mention two differences or ways of being situated in Eastern Europe specific to STS. The first is connected to something I call "the disappearance of the Second World," while the second to the asymmetries in knowledge production, especially in a peripheral or semi-peripheral situation. Together with Krzysztof Abriszewski and Michał Wróblewski,[1] we call the latter "suitcase science." Both differences seek a stabilizing ideology at the intersection of hegemonic economic order and the production of knowledge.

The Disappearance of the Second World

Let us begin with the first, the mysterious disappearance of "the Second World." Before 1989, the Cold War architecture of the world divided into three parts: the First (capitalist), the Second (socialist) and the Third (developing) World. As a historical consequence, semi-peripheral Eastern Europe belonged the Second

---

1   Andrzej W. Nowak, Krzysztof, Abriszewski and Michał Wróblewski, *Czyje lęki? Czyja nauka? Struktury wiedzy wobec kontrowersji naukowo-społecznych* (Poznań: Wydawnictwo Naukowe UAM, 2016).

World. Its position stretched between a frequently imitating attitude towards the core countries of the modern world-system, and the seeking of alternative routes of development and supporting liberation movements in the Third World. After 1989, the tripartite division of the world was substituted for a bipartite relation between the North and the South. The challenge for the semi-periphery and its knowledge production – in the context of STS – lies in finding strategies of development within this division.

It is worth discussing the metaphor of "the disappearance of the Second World"[2] by referring to the movie *Django Unchained* directed by Quentin Tarantino (2012). The Western movie played in pre-Civil War USA features a couple of mutually entangled protagonists: the German dentist, Dr. King Schultz (brilliantly played by Christoph Waltz) and the eponymous Django, the black slave he liberated (Jamie Foxx). Their relationship is intriguing. Django owes his liberty to dr. Shultz. However, the white liberator is not an abolitionist or an idealist – he is a pragmatist, a bounty hunter following the dangerous Brittle brothers, and he knows that only Django can lead him to them. Schultz, outlined stereotypically like the "German" characters in Karl May's books, buys out Django and promises him freedom after the Brittle brothers are captured. But when this happens, the relationship between our characters does not end. Schultz and the liberated Django keep working together, looking for wanted criminals. Schultz earns his living, and Django plots to find and rescue his wife, whom he had lost at a slave market in the past. At the same time, the alliance between the characters is cemented by the fact that they liberate slaves. At last, their efforts are rewarded: they find out where to find Django's wife. She belongs to a slave merchant known for organizing cruel competitions: death duels between slaves. In order to fulfill Django's goal to liberate his wife, our characters have to face this powerful enemy. They plan an intricate game: Dr. Schultz pretends to be a naïve gambler interested in the practice of slave fights, while Django plays the role of his personal bodyguard and duel expert. However, their precautious plan ultimately fails and crumbles into bloody shambles. But finally, Django wins and liberates his wife. Unfortunately, Doctor Schultz dies.

However, what is key for my considerations here is not the evil Master, but the character of Stephen, an old, trusted black slave and majordomo at the slave owner's mansion. He is the one who betrays our characters as the most loyal

---

2 Andrzej W. Nowak, "Tajemnicze zniknięcie Drugiego Świata. O trudnym losie półperyferii," in: *Polska jako peryferie*, ed. Tomasz Zarycki (Warszawa: Wydawnictwo Naukowe Scholar 2016), pp. 86–104.

servant of the cruel Master. In fact, he is the pillar of the system: his solidarity with the tormentor rather than with the victims maintains the structures of oppression. The majordomo remains faithful to the slave structure of power until the end. On the one hand, it deprived him of his freedom, but on the other, it gave him status, provided him with meaning, placed him clearly within the hierarchy of power and merits. When he dies during the final struggle, the viewer can feel cathartic satisfaction no lesser than in the case of the earlier death of the main adversary.

Django serves me as a metaphor of the semi-peripheral identity of the Second World. I am particularly interested in the struggles of the countries that find themselves in this strange position in the global system of the distribution of power and recognition. This particular position requires a constant struggle for being provided with some form, which sets it apart from the extreme constituents: the First and the Third World. These identities and roles of the Master and the Servant/Slave are clearly defined and historically ascribed. The struggle concerns either the liberation (the Third World) or the maintaining of the privileged position (the First World). But in the case of the Second World, the situation is paradoxical: the intermediary regions need to assure themselves with identity, fight for it, create it, and simultaneously liberate themselves from it.

Based on the above, I narrow down my analysis of the Second World to presenting the situation of Eastern Europe. Before 1989, as a semi-peripheral region belonging to the Second World, the region resembled the story of Dr. Schultz. In relation with the Third World, communist Eastern Europe took on a paternalist attitude by "taking care" of developing countries. Analogically to the film, this "liberation" or help was connected with mutual business dealings. Each side had its own aims and carried them out within this paternalistic brotherhood. This also connected to the transfer of technology, educating people from the Third World in the countries of Eastern Europe, investing in Third World countries or building factories, schools, hospitals, etc. there. This relationship influenced an attempt to (re)define the center-periphery system.

The identity of the semi-peripheral Eastern Europe was stabilized by its belonging to the so-called Second World under the hegemony of the Soviet Union. Due to its particular location and relative development, it made possible for its satellites to have relative latitude by creating pressure on hegemonic Western powers. In Immanuel Wallerstein's scheme, the Soviet Union was a semi-periphery that tried to construe an alternative center of the world system through the bipolar world of the Cold War. Russian modernization aimed to achieve a hegemonic position in the modern capitalist world-system, but although the Soviet Union could not manage to fully achieve a status as a core or central country, the existence of an alternative represented by the relatively developed Eastern Bloc changed the geopolitical, economic, and ideological landscape for decades. As our dentist from the movie, the Second World countries could, by participating in local ideological, political, and military conflicts and in scientific and technological exchanges, on the one hand, liberate Third World countries from colonialism and imperialism, and, on the other, legitimize their own policies as progressive. Pragmatism and idealism were entangled in an inseparable knot. The Soviet Union, as well as the entire Eastern Bloc, was dependent on ideological legitimization. It was necessary to maintain both the internal order and, more interestingly for us, the system of international relations. In the Soviet Union, regardless of actual deficiencies and shortages[3] in the field of infrastructure, the reference to science, technology, and progress played an important role in shaping the architecture of ideas and the transfer of knowledge and technology during the Cold War. Although all this ideological configuration ended in 1989–1991, but unfortunately in a manner different from the movie.

Firstly, the ideological antagonist symbolized by the slave plantation owner in the movie – the capitalist world – did not fall, and the imperial and neocolonial politics are still doing well. Secondly, the Third World countries – Django – did not achieve their goals, did not liberate themselves from their peripheral position, although interestingly the position of India and China has been rising in recent years.[4] But what about the Second World of Eastern Europe? Well, if the story had happened the way it did in the movie, then the Second World would die, and owing to that, the slave order would fall, and our liberated slave Django – the Third World – would liberate himself and the others through a world revolution. However, we know that this did not happen – history followed a different script. Despite this, after 1989 the Second World disappeared. When we talk today about ideological legitimization of global inequalities, the two-pole language dominates, describing the sides of the system as the rich North and the impoverished South. There is no space for other ideological actors. To return for a moment to my metaphors connected to the movie, we have to state that the Second World, once the Soviet Union had been defeated, changed its role (this ideological role started to corrode even earlier before 1989). From

---

3   Kate Brown, Plutopia, *Nuclear families, atomic cities, and the great Soviet and American plutonium disasters* (Oxford: Oxford University Press, 2013).
4   Andre Gunder Frank, *ReOrient: Global Economy in the Asian age* (Berkeley: University of California Press, 1998).

a solidary and paternalistic partner of the Third World it transformed into Stephen, a slave majordomo who stabilizes the system, sucks up to the Master, counting on receiving at least some crumbs from the master's table. This shift had grave consequences for the entire global system. Not only did the Second World lose its relative self-reliance, but it also left the scene and together with it there disappeared an alternative source of creating history, producing knowledge, stabilizing the ideology of progress. The Second World as a complementary element of a trialectic system from the times of the Cold War disappeared from the ideological landscape of geoculture. When this construction ceased to exist, the Second World lost its already wobbly identity. In most cases it became only an adulatory element of the two-pole system or – as in the case of Ukraine – it tumbled to the level of periphery. This resulted in the First World losing its ideological and political antagonist, and the Third World, its paternalistic partner. The world became ideologically flatter but simultaneously politically and ontologically even more complex, with all the consequences of this state. Technology, science, the development of knowledge, which at the time of socialism could expand within a particular and partial autonomy-autarky were hastily incorporated into the architecture of the modern capitalist world-system.[5]

Starting a narrative by referring to a popular movie resembles the style of the Slovenian philosopher, Slavoj Žižek; but it also opens the question regarding the strategies of representation and gaining a voice by philosophers (here in our case STS scholars) from postsocialist Eastern or East Central Europe in countries of the center. Žižek chose strategies of self-exoticization, playing games, and clowning around. This is an ambiguous strategy: a Hollywood movie becomes a source of metaphors to speak about semi-peripheral identity. And yet, the USA movie industry is, next to denim jeans, one of the most distinct symbols of the imperial cultural hegemony of the USA, which is still the dominant center of the modern world-system. For theoreticians, scholars, and artists from semi-peripheral countries, joining the center is always an ambiguous venture stretched between the risk of assimilation to the point of dissolving into the hegemonic narrative and losing one's own voice. Other strategies include mimicry or essentializing local identity, which is visible in contemporary Poland, where popular journalist and right-wing activist Wojciech Cejrowski in Polish television dressed as a US cowboy accompanied by country music was promoting essentialized nationalistic historical view of Polish history and trying to positively revaluate

---

5   Immanuel Wallerstein, "Social science and the communist interlude, or interpretations of contemporary history," *Polish Sociological Review*: 117/1997, pp. 3–12.

the obscurant backwoods tradition called in Polish "Ciemnogród" (Dunceville). Markedly, the term "Ciemnogród" refers to a novel written by Stanisław Kostka Potocki *Podróż do Ciemnogrodu* (Journey to Dunceville) written in 1820, which was an Englithment critique of backward and ultraconservative Poland.[6]

Although strategies of "separateness" emphasize the politics of local identity, they often result in essentialization, with the additional danger of succumbing to autarky and, as a result, the impossibility to articulate one's own voice. Each of these strategies has its advantages and drawbacks. It is clearly visible that inventing and looking for an autonomous form remains the current task of the semi-peripheral voice.[7]

The countries of the center have it easier: it is possible for them to define themselves through the figure of the Master or through a game contesting this figure. Similarly, the peripheries are defined so strongly that they know what to fight against. The semi-peripheral identity is spread out and torn between two ordering principles: the dominating narrative of the center and the potential voices of either resistance or succumbing from the peripheries. Today it is difficult for the countries of the former Eastern Bloc to seek identity as they lost their relative dependence from capitalist center even it was paid for dependency form Soviet Union and, to a large degree, have systemically joined the West, becoming its less privileged part. They are unable to show, at least at the moment, "a potential alternative," but instead have become the internal peripheries of the capitalist center. The Second World lost the ideological capacity to produce its own narrative along which to develop its identity. Before 1989, a vision of an alternative route of modernization provided this. But today, the countries of the former Second World are unable to, at least for now, propose their own narrative. The relatively closest reference point is Russia, which is capable of outlining proposals for itself and its neighbors, suggesting a Euro-Asian ideology.[8] However, it is clearly visible that these projects do not have such scope and structure-creating power as the universalist communist ideology promoted by the Soviet Union had. The disappearance of the Second World meant an end to an alternative in the fight for who defines the course of history and who promotes better universalist ideologies and superior scientific and technological development. The fall of the Soviet Union and the dismantling of the Second

---

6 Czesław Miłosz, *The History of Polish Literature, Updated Edition* (Berkeley: University of California Press 1983), p. 204.
7 Witold Gombrowicz, *Ferdydurke* (New Haven: Yale University Press 2012).
8 Alexander Dugin, *The Fourth Political Theory* (Budapest: Arktos Media Ltd. 2021).

World resulted in not only a crisis of leftist ideas, but above all, a serious crisis of universalist ideologies symbolized by the United Nations and the idea of human rights. During the Cold War, these latter were used instrumentally by both sides of the conflict and gained significance by becoming a field of rivalry. The disappearance of the Second World shattered this construction – in a world divided into the North and the South there are no mechanisms stabilizing universalist ideas like it was Second World communist alternative, and we are left asking what semi-peripheral modernity was, what it is and what future awaits it.

Suitcase Science

Now we may move to the second difference – or challenge – of "suitcase science."[9] The structural pressure of the global system on ways of practicing science (including STS) is evident in that in the postsocialist Eastern European region the greatest scientific advancement comes from practicing imitative science ("science in a suitcase"), which in turn disrupts the harmonious blood circulation of knowledge within a scientific field. By "suitcase science" we understand a situation in which it is the easiest to pursue an academic career (in Eastern or East Central Europe) through "bringing" ready-made concepts from countries of the center and implementing them locally through imitation. The metaphor of a "suitcase" derives from observing Polish academia in the 1990s, when many academic careers were closely determined by literally bringing books and xerox copies from Western universities to Poland. In light of this, to answer the question pertaining to STS and asked in the spirit of STS about the specific, situated ways of practicing science in Eastern or East Central Europe, together with Abriszewski and Wróblewski we modified Latour's circulatory model of scientific practice.[10] Our aim was to break down the "common sense liberalism" dominating within technoscience by putting our considerations within the framework of the concept of Immanuel Wallerstein's capitalist world-system.

The well-known circulatory model, presented by Latour and based on the historical case of research conducted in France before WWII on atomic energy, presents scientific theory as a heart of a networked vascular system, which can beat and develop only when its whole system is circulating and developing. This

---

9 Krzysztof Abriszewski, "Nauka w walizce," *Zagadnienia Naukoznawstwa*, Vol. 52, 1(207)//2016, pp. 87–112; Nowak, Abriszewski and Wróblewski, *Czyje lęki? Czyja nauka?*.
10 Bruno Latour, *Pandora's Hope: Essays on the reality of science studies* (Cambridge Mass.: Harvard University Press, 1999).

system consists of four loops: the mobilization of the world, autonomization, alliances and public representation. The aim of this model is to present the circulation of knowledge within a society. But Latour didnot really developed this model to embrace the global circulation of knowledge and its inequalities. We broaden Latour's model by asking: what does the circulation of science look like when it is based on "a suitcase" and not on a laboratory?

Let us now simply superimpose this onto Latour's system of four loops:

1. The range of *mobilization* of the world is ruined; academics or scientists do not conduct research, but merely copy someone else's research. It is a small loop.
2. There is also a problem with *autonomization*. Reproducing "Western" texts means little exchange and lack of building actual thought collectives. Researchers are not oriented towards each other, but towards the center.
3. The list of *alliances* is the most visible example of neocolonial dependency. "Suitcase academics" receive grants and rewards for academic achievements, working at the top universities in the world, but also at extra-academic institutions. However, all this is based on mimicry.
4. Within *public representation*, suitcase science is based on "reflecting the light" of the theories of the center. Most importantly, this does not concern its contents, but rather the fact whether it can provide the "Author" with appropriate status as a representative of the Centre (the West) "in touch" therewith.[11]

It turns out that, in contrast to Latour's account, in semi-peripheral science the heart in the center of a circulatory system is not the laboratory or theory, but a career-oriented academic bringing books in their suitcase. To use a metaphor suggested by Krzysztof Abriszewski, we are dealing with an undead heart, neither alive nor dead; we are dealing with zombie science.

Following from the above, my reply to the question regarding the situation of STS in Eastern or East Central Europe needs to be quite pessimistic. There is a need for a change at the political, economic, and institutional level. Without redefining the position of the region, inquiring about the possibility of a more autonomous position it would be impossible to defeat "suitcase science." Without this, most efforts will be made not in producing knowledge, but in making careers by means of imported knowledge, thus reproducing hegemonic

---

11 Cleaver, Harry, "On schoolwork and the struggle against it." in: "Unpublished Notes." 2003. https://la.utexas.edu/users/hcleaver/OnSchoolwork200606.pdf, (2 April 2022).

relations. In this approach, STS becomes only one of currencies thanks to which successes in academic field can be achieved. The question about the possibility of practicing STS in Eastern or East Central Europe should thus be: is it possible to reclaim those promises which were stabilized by the Second World but in new ideological configuration?

Bibliography

Abriszewski, Krzysztof. "Nauka w walizce." *Zagadnienia Naukoznawstwa*, Vol. 52, 1 (207)/2016, pp. 87–112.

Brown, Kate. *Plutopia: Nuclear families, atomic cities, and the great Soviet and American plutonium disasters*. Oxford: Oxford University Press, 2013.

Cleaver, Harry, "On schoolwork and the struggle against it." In: "Unpublished Notes." 2003. https://la.utexas.edu/users/hcleaver/OnSchoolwork200606.pdf, (2 April 2022).

Dugin, Alexander. *The Fourth Political Theory*. Trans. Mark Sleboda and Michael Millerman Budapest: Arktos Media Ltd., 2012.

Frank, Andre Gunder. *ReOrient: Global Economy in the Asian age*. Berkeley: University of California Press 1998.

Gombrowicz, Witold. *Ferdydurke*. Trans. Danuta Borchardt. New Haven: Yale University Press, 2012.

Latour, Bruno. *Pandora's Hope: Essays on the reality of science studies*. Cambridge Mass.: Harvard University Press, 1999.

Miłosz, Czesław. *The History of Polish Literature, Updated Edition*. Berkeley: University of California Press, 1983.

Nowak, Andrzej W. "Tajemnicze zniknięcie Drugiego Świata. O trudnym losie półperyferii." In: *Polska jako peryferie*, ed. Tomasz Zarycki. Warszawa: Wydawnictwo Naukowe Scholar 2016, pp. 86–104.

Nowak, Andrzej Wojciech, Krzysztof Abriszewski and Michał Wróblewski. *Czyje lęki? Czyja nauka? Struktury wiedzy wobec kontrowersji naukowo-społecznych*. Poznań: Wydawnictwo Naukowe UAM, 2016.

Wallerstein, Immanuel, "Social science and the communist interlude, or interpretations of contemporary history." *Polish Sociological Review* 117/1997, pp. 3–12.

Krzysztof Abriszewski
Translated by Maciej Smoczyński

What Does Theory Do in the Humanities?

Abstract: Following the two-faced model of science from Bruno Latour's *Science In Action* (1987), in which well-known ready-made science is opposed to science in action, I propose a similar approach to theory in the humanities. Theory, rather than being a ready-made product of scientific conduct, may be viewed as a process itself being part of other processes. The theory-in-action approach is argued against common theory-as-a-language-structure approach (i.e. logical structures of propositions). It is argued that empirical examples of theoretical investigations and arguments in humanities clearly are not stable theoretical structures, but processes and actions. The main part of the text argues for several such actions: see something else, forget your work, circulate, engage emotions, rebuild the world.

Keywords: theory in action, humanities, Actor-Network Theory, orientalism, criterial prefocusing

1. Theory and Science in Action

In his *Science in Action* Bruno Latour[1] shed light on two faces of science evoking Roman god Janus to elaborate his point. Ready-made science was the face that was better known and more thoroughly studied. The other one, which Latour named science in action, is the face usually overlooked both by scholars studying science and wider, non-scientific audience. Latour argued for two face model of science in order to expose and study its huge part. What I am proposing in this paper is to extend his argument to theory in humanities and social sciences, and instead of studying ready-made face of theory, pay more attention to its "in action" face. In other words instead of investigating the construction of theory in humanities, let us pose the following question: what do theories do? Theories, which we construct, operate with and use to analyse and criticise in social and human sciences? As already mentioned, the traditional interest in theory in philosophy of science and scientific methodology pointed in another direction. A question rather used to be asked what theory is, what a good theory

---

1 Bruno Latour, *Science in Action. How to Follow Scientists and Engineers through Society* (Cambridge Mass.: Harvard University Press, 1987).

is characterised with.² All those considerations tacitly assumed that theory is a culminating moment in science: it was new knowledge that all investigating actions led to.

In the classic textbook by Stefan Nowak, *Methodology of Sociological Research. General Problems*, alongside such chapters as "Kinds of propositions" or "Substantiation of statements. Empirical verification of Hypotheses" one can also find Chapter 7 "Construction of theories," where Nowak writes the following:

> By theories we shall mean *sets of laws arranged so as to form a certain internally coherent logical structure*, regardless of whether the propositions in those theories concern phenomena which are directly observable in greater or lesser measure. However, not every internally ordered set of general propositions is worth calling a theory and not every mode of systematizing propositions worth calling a theoretical systematization.³

And further:

> we shall make reference to *theoretical systematization* whenever the sets of scientific propositions are arranged so as to allow use of the complete sets in explaining and predicting events, or to enable some laws to be explained in terms of other laws or, again, to derive new laws from these already accepted in the given science. Thus, a *theory is a system of laws*, in which the relations manifested in it between the individual laws either enhance the utility of those laws in explaining and predicting events or else enable some laws to be explained in terms of others, and make it possible to deduce (predict) new laws from the set of recognized laws.⁴

But do humanities scholars dealing with theory really work this way? Do they really construct structures of laws, propositions and sets?⁵ As a telling

---

2   See for example: Adam Grobler, *Prawda i racjonalność naukowa* (Kraków: inter esse, 1993), Adam Grobler, *Metodologia nauk*, (Kraków: Wydawnictwo Aureus, Wydawnictwo Znak, 2006), Jerzy Kmita, *Essays on the Theory of Scientific Cognition* (Dordrecht: Springer-Science+Media, 1991), Stefan Nowak, *Methodology of Sociological Research. General Problems* (Dordercht/Boston: D. Reidel Publishing Company Warszawa: PWN – Polish Scientific Publishers, 1977), Wojciech Sady, *Spór o racjonalność naukową. Od Poincarégo do Laudana* (Wrocław: FUNNA Sp. Z o.o. – Wydawnictwo, 2000), Frederick Suppe, *The Semantic Conception of Theories and Scientific Realism* (Chicago: University of Illinois Press, 1989).
3   Nowak, *Methodology of Sociological Research*, p. 360.
4   Nowak, *Methodology of Sociological Research*, p. 360.
5   On the other hand, Charles Taylor, exploring the interrelation of practice and theory, notes: "I want to maintain that gaining clarity about the practice of theorizing will help us to understand more about the scope and validity of our theories. Being more reflectively clear about what we do in our theoretical activity will help us to answer

counterexample let us read randomly chosen fragments from a book to which I will also refer further in this text, from *Orientalism* by Edward Said:

> This attitude is profoundly anti-empirical. And so, indeed, is the Orientalist attitude in general. It shares with magic and with mythology the self-containing, self-reinforcing character of a closed system, in which objects are what they are because they are what they are, for once, for all time, for ontological reasons that no empirical material can either dislodge or alter. The European encounter with the Orient, and specifically with Islam, strengthened this system of representing the Orient and, as has been suggested by Henri Pirenne, turned Islam into the very epitome of an outsider against which the whole of European civilization from the Middle Ages on was founded.[6]

Further in the book we read:

> Rather than listing all the figures of speech associated with the Orient—its strangeness, its difference, its exotic sensuousness, and so forth—we can generalize about them as they were handed down through the Renaissance. They are all declarative and self-evident; the tense they employ is the timeless eternal; they convey an impression of repetition and strength; they are always symmetrical to, and yet diametrically inferior to, a European equivalent, which is sometimes specified, sometimes not.[7]

Then, in the beginning of the next chapter Said notes:

> *Orientalism* is the generic term that I have been employing to describe the Western approach to the Orient; Orientalism is the discipline by which the Orient was (and is) approached systematically, as a topic of learning, discovery, and practice. But in addition I have been using the word to designate that collection of dreams, images, and vocabularies available to anyone who has tried to talk about what lies east of the dividing line.[8]

Said's argument is based largely on analysis of literature, so it contains analysis of empirical material of a sort, but even if we pick any of hundreds of (purely) theoretical works of humanities and social sciences, there is only a marginal chance to find a sequence of propositions, that form sets and structures resulting in laws and predictions. This does not mean that there will not be any, just as it does not mean that the stance represented by S. Nowak is simply wrong. Let us note that these elements appear in Said's argument, in the last quotation we can find if not a full blown theoretical structure of orientalism, then at least some propositions.

---

    questions which we cannot even properly pose as long as we remain convinced that social theory is a straightforward matter of designing hypotheses and comparing them to the facts" (Taylor 1985: 91).
6 Edward Said, *Orientalism* (London: Penguin Books, 2003), p. 70.
7 Said, *Orientalism*, p. 72.
8 Said, *Orientalism*, p. 73.

However, in Said's work, like in other theoretical works of human and social sciences, we do notice something else. Among this something else sets of propositions can occur. Let us focus on the way the argument in *Orientalism* meanders. In the first of the quoted paragraphs "the Orientalist attitude" is related to the general idea of magic and religion, which in turn are related to a still greater abstraction of "the self-containing, self-reinforcing character of a closed system," to proceed to European confrontation with the Orient, and next make a leap to Henri Pirenne. In the second paragraph Said smoothly passes from the abstraction of "the figures of speech associated with the Orient" to history ("the Renaissance's" legacy), the grammatical tense changes to "the timeless eternal." Finally, in the third quotation we obtain elements of a definition, but its structure is more reminiscent of Baroque strategy than more minimalist efforts from Nowak's book.

We have, then, two forms of ordering, or two styles. The order proposed by S. Nowak in his textbook and Said's order of a living text. While the first is consistent with the way science is "officially" perceived,[9] one can hardly reject the other as a whim, or an exception. On the one hand, Said is a typical case, to a degree, but on the other hand, as an original thinker, he counts among the most important inspirators of a whole research area known as "postcolonial studies."[10]

---

9 See for example the definition of theory in the accessible online *Encyklopedia PWN* (*Polish Scientific Publishers Encyclopaedia*): "scientific theory: a methodological system of logically and thematically ordered statements, linked with specific logical relations, existing in a specific science and fulfilling criteria of scientific rigour and methodological correctness proper to it. In different sciences scientific theory has different structures, different levels of justifiability and internal consistency; in formal sciences (mathematics and logic) the name of scientific theory describes a system of theorems consisting of a set of axioms and logical consequences (deductive system, formalised system); in empirical sciences a scientific theory of a given area of phenomena is either a set of sentences explaining given phenomena, which, combined with scientific laws explaining them and logical conjectures, forms a unit of scientific knowledge about the given domain, or any system of statements concerning a given area and containing theoretical terms (referring to objects or definitions impossible to perceive directly), or otherwise a system of scientific laws, general conjectures and definitions linked logically and thematically so that from the most general of them it is possible to infer all less general statements, and expressed in a way that allows empirical control (verification); the task of a scientific theory in empirical sciences is not only explaining facts, but also predicting them;" *Encyklopedia PWN* (*Polish Scientific Publishers Encyclopaedia*) (9 Sep. 2021) http://encyklopedia.pwn.pl/haslo/teoria-naukowa;3986514.html.

10 See for example: Leela Gandhi, *Postcolonial Theory: A critical introduction* (Crows Nest: Allen & Unwin, 1998), pp. 64–80, Ania Loomba, *Colonialism / Postcolonialism*

So it would be difficult to attack Said for not adhering to "proper" methodology without attacking the whole research field.[11]

Not accidentally I called Said's text baroque, having in mind its intrinsic complexity, and maybe even (numerous) complexities. In my scarce commentary I remarked on the ontological transitions through which the text leads the reader: one abstraction level changes to another, hypothetical entities stand side by side with real ones, events and beings intermingle, and so do authors and their texts, its contents and interpretations. The Actor-Network Theory literature assumes that scientific texts are by nature "constitutively heterogeneous."[12] A scientific text must face several problems, among them the absence of the author, who could defend it, answering charges, as well as the reader's boredom, fatigue and resentment. And these are not all.

It is reasonable to suppose that Nowak's text, as many of a kind, presents an ideal, which is known to be forged after the model observed in physics and, partially, mathematics. In contrast to this, I reckon that in Said's book we can see how theoretical work in humanities actually looks like. The vast majority of theoretical texts in the area of human and social sciences do not form a quasi-formal structures of propositions, and the vast majority of their readers expects reading a continuous, literary text, as opposed to analysis of some formal, or quasi-formal, structure. Neither the author of theoretical works in humanities, nor their recipient focuses their analytical work on theory construed this way. On the other hand, though, I do not wish to slip into some kind of (vulgar) post-modern argument, according to which theoretical structures in the vein of Nowak would only be a scientistic fantasy and consequently humanities were to be treated just as a form of writing.

(London and New York: Routlege, 2015). In her *Postcolonial Theory*, Leela Gandhi writes: "Commonly regarded as the catalyst and reference point for postcolonialism, *Orientalism* represents the first phase of postcolonial theory" (Gandhi 1998: 64).

11 As we know, attacks of this kind occur from time to time, anyway. Among the most well-known are attacks on psychoanalysis as an unscientific method and science wars, with the famous critique by A. Sokal and J. Bricmont, and the former's previous hoax.

12 John Law writes: "Texts in science are *constitutively heterogeneous* in nature and scientists operate not only upon the 'internal' scientific world but simultaneously upon 'external' as well. This is because the strength of an argument, whether scientific or not, rests upon the number and power of the elements that may be borrowed to give it force." John Law, "The Heterogeneity of Texts," in: *Mapping the Dynamics of Science and Technology*, eds. Michel Callon, John Law and Arie Rip (London: The MacMillan Press Ltd. 1986), p. 79.

Even if we thought what is depicted by S. Nowak as fiction, then it is not in the vein of fiction as a tall story, but rather as a Kantian regulative idea or a certain ideal, fiction as a stable substantial component of cultural reality. The justification of such a distinction is provided by my adoption of what ANT says about scientific texts. They are multilayer constructions, heterogeneous by nature, trying to combine different resources, entities, places and times. They strive (again, in their ideal form) to lead the reader from what is known and what is easy to agree with to the unknown, unheard of and, at the first glance, unbelievable. Readers must follow the whole route not getting lost, as then the work would be futile.[13] The purpose of the text – and that may be the key difference between what I propose and the "humanities as a form of writing" approach – is not to entrance the reader, or to induce fascination in the text as such. It is meant for the reader who has read the text to subsequently go further, investigate something more and something else. A scientific text is therefore like Wittgenstein's ladder, which is used to reach the next level and then abandoned.

Having explained why disagreement with Nowak's approach (and similar ones) does not mean falling into anti-scientific position, I would like to show the basic mechanism of research perspective change that I am suggesting here in contrast to the one represented by Nowak's book. This mechanism may be called the "STS switch," in reference to the area that I reach for,[14] the "verb switch," hinting at Ryszard Nycz's "culture as a verb" concept,[15] or simply the "processual switch." It is very simple: Nowak's approach stands on the side of "ready-made science." When research is completed, debates lose their fervour, and their results find their way to textbooks which the next generation of researchers is

---

13 See for example Krzysztof Abriszewski, *Wszystko otwarte na nowo. Teoria Aktora-Sieci i filozofia kultury* (Toruń: Wydawnictwo Naukowe UMK, 2010), pp. 27–42, Latour, *Science In Action*, pp. 21–62, Bruno Latour and Françoise Bastide, "Writing Science – Fact and Fiction: The Analysis of the Process of Reality Construction Through the Application of Socio-Semiotic Methods to Scientific Texts," in: *Mapping the Dynamics of Science and Technology*, eds. Michel Callon, John Law and Arie Rip (London: The MacMillan Press Ltd. 1986), pp. 51–66, Bruno Latour and Steve Woolgar, *Laboratory Life: the Social Construction of Scientific Facts* (Los Angeles, London: Sage, 1979), pp. 151–186.

14 Latour, *Science In Action*.

15 Ryszard Nycz, *Kultura jako czasownik. Sondowanie nowej humanistyki* (Warszawa: Instytut Badań Literackich, 2017). In the same way as Nycz, who speaks about culture as a verb, we could speak here about "science as a verb" or "theory as a verb." These are only specific applications of the thinker's more general formula to selected areas of culture or its artefacts.

going to learn from, then a theory, purified of any background noise becomes culminating point of scientific effort. However, if we note that the pre-requisite for this is the closure of a certain historical sequence, we may acquire interest in something else: science in its "hot phase," where disputes are on, scientists struggle for research funds, results are unsure or not yet known, and even it is not utterly clear which the studied problem is. Science at this stage of uncertainty is "science in action." As mentioned earlier, I refer here to Latour's classic *Science In Action*.[16] The duality is represented there by the figure of Roman deity Janus. Science, just like him, has got two faces, old and young. The old one symbolises ready-made science, the young one stands for science in action, riddled with uncertainty.

Consequently, moving from interest in construction of a theory in the vein of Stefan Nowak to the question what a theory does, I employ such a processual switch, which, having recognised this duality of science, allows now to establish the range of interest for the study of science. Using the switch lets us make a passage from the old face, which is ready-made science, in which theory is already constructed and it has a determined structure, possible to recognise, to the young face, that is, science at work, in which theory gets messed up with many other elements, but does not provide a conclusion for the whole process, as the process has not finished yet, it is going on. I maintain that at this stage theory and theoretical work participate in pushing the whole task forward. However, using the processual switch also enforces changing the answer to the question what the process is and how long it lasts. It certainly cannot be reduced to the simple sequence saying "researcher > research > result > final theory." Instead, I propose adopting another formula, more in the ANT spirit, stating that the laboratory is a place where innovations are forged, and it lasts much longer than the former sequence suggests. To a degree, the further part of this text strives to capture this very process in the form of an answer to the question what theory does. The transition from ready-made science to science in action is equivalent, as the switch's name says, to abandoning the static approach in favour of following the processes, thus I am going to describe the actions leading to forming a theory in a few points: distinguished moments in the sequence of events.

---

16 Latour, *Science In Action*.

2. Action 1: See Something Else

In the first episode of the first series about the world's most famous detective Sherlock Holmes, that transferred his adventures into early the twenty-first century reality,[17] the creators adopted an interesting communication convention. In the scenes where Holmes performs his famous deduction, that is proceeds from observations to conclusions, double communication coding was applied. Thanks to this the viewer is able to simultaneously follow the "standard" course of events and the detective's reasoning. The same can obviously be achieved through presenting the events first, and then the explanations as in Conan Doyle's novels and stories. Nevertheless, the chosen method allowed the viewer to see something else: the difference between the "normal" perspective and Holmes's shrewd insight. This double coding is very simple: the film's action goes on as usual, we can see people who investigate the crime scene, talk to one another, but this image was additionally appended with written information about what the famous detective is able to notice and deduce. For example, one of the takes shows a close-up of the victim's hand, on which we can see three fingers and a ring on one of them. The additional message, which is information of what only Holmes notices, says "unhappily married 10+ years."

This way the series interestingly shifts the emphasis regarding the detective's cognitive procedure. In the literary version the stress was on the reasoning, deduction, making subsequent steps proceeding from the obvious and visible to anyone, to the inadvertent, surprising, inferred. The medium of film, thanks to the imminence of a visual message, depicts the moment of recognition, perception, noticing something else.

In this perspective *Sherlock* provides an excellent depiction of what theory does: namely, it lets us see something else or just see something more. It is theory's first action to which I would like to draw attention. If a text successfully conveys a theory, then the readers not only (or not exactly) is able to repeat what they have read, as well as summarise, remember or retell it, but also will perceive the world in a different way, will see other things. How theory has furnished the

---

17 I am talking about *Sherlock*, whose first season was aired in 2010. Its main creators are Steven Moffatt and Mark Gatiss. The episode which provides me with the detailed example, the first of the first season, is titled "The Study in Pink" and it was directed by Paul McGuigan, see (9 Sep 2021) https://en.wikipedia.org/wiki/Sherlock_%28TV_series%29.

world is going to filter their perception. The theory's ontology is going to become the reader's epistemology.[18]

To illustrate this, let us recall an example referring to a certain socio-scientific controversy, where this moment of altered view became a flashpoint and an important element later.[19] Let us return to the already quoted book by Edward Said and read another passage:

> But there were other uses for latent Orientalism. If that group of ideas allowed one to separate Orientals from advanced, civilizing powers, and if the "classical" Orient served to justify both the Orientalist and his disregard of modern Orientals, latent Orientalism also encouraged a peculiarly (not to say invidiously) male conception of the world. … / The Oriental male was considered in isolation from the total community in which he lived and which many Orientalists, following Lane, have viewed with something resembling contempt and fear. Orientalism itself, furthermore, was an exclusively male province; like so many professional guilds during the modern period, it viewed itself and its subject matter with sexist blinders. This is especially evident in the writing of travelers and novelists: women are usually the *creatures of a male power-fantasy*. They *express unlimited sensuality*, they are more or less *stupid*, and above all they are *willing*.[20] (emphasis – K. A.)

Said's classic work describes a certain conglomerate of phenomena, which gains the name of "Orientalism." If the reader internalises Said's reasoning, forming one cognitive perspective, then she is likely to recognise cases of Orientalism in her environment. For example, reading the following piece of travel literature:

> I have always thought that there is a certain regularity in *white men's desire* to possess black women. I do not think, however, it were about sexologists' and historians' favourite motive of domination. I rather see this force as an atavism. As a sublimation of the desire to fornicate with animals. So felt the outstanding French artist Jean-Paul Goude. His Negresses in computer photos were mares, caged predators. Black girls in Africa are almost always wild. It is a result of being raised in a society built on a male pack leader.

---

18 Mariola Kuszyk-Bytniewska investigates in details and argues for the significance of the shift of perspective in social sciences from the epistemocentric one to onto-epistemological one. Mariola Kuszyk-Bytniewska, *Działanie wobec rzeczywistości. Projekt onto-epistemologii społecznej* (Lublin: Wydawnictwo Uniwersytetu Marii Curie-Skłodowskiej, 2015).
19 For drawing my attention to the controversy and debates on it, I am grateful to Monika Bobako and Andrzej W. Nowak. Note that Said's name, whose work I use here, does not appear in the controversy itself. It is rather an element of context, that is postcolonial studies, and which I bring to light trying to reconstruct the mechanisms that are not explicitly expressed in the controversy.
20 Said, *Orientalism*, p. 207.

So live lions and gorillas, so do tribes of black Africa. Their girls have distrust inscribed in their genes. They are skittish like small antelopes, while they take their beauty after big cats. They are not suitable for domestication. They speak a different language even when they use the same words as us. Most often, though, they do not speak in the presence of men. They nuzzle up to them. Or they hunt, like lionesses. African prostitution is this kind of hunting. For most Africans sex is life's only pleasure. They use it in a way just as natural and spontaneous as animals. If they do not work and do not make love, boredom kills them. Their prostitution is not consequent. They take money when they need it. Sometimes they only ask for breakfast. *They frequently simply ask to be possessed. And they are unimaginably beautiful. I do not think there is any healthy man capable of resisting their charm.* For it is not human charm, to which we are accustomed, but animal charm. These girls ask to be covered. They begin then to smell differently, suddenly, like halved fruit. *Everything in them speaks about intercourse, eyes, lips, hands.* At last they talk as cat females would speak if they could. I met girls like this later in Florida 2000 clubs in Nairobi and in the streets of Mombasa, and repeatedly in Uganda. They had run away from their tribes to big cities or had already been born there and *they want to live to the rhythm of their pulsating bodies.* Maybe they want to relinquish at least a part of the joy which a thousand-year-old ritual of female initiation takes away from them.[21] (emphasis – K. A.)

A case of such an interpretation, i.e. seeing something else, lay at the basis of a modest controversy that sparked in Poland in February 2011 and took place mainly on the Internet, only slightly reflected in official, paper press, one over the book *Chwila przed zmierzchem* (The Moment Before Dusk) by a radio host known from Polish Radio's 3, published in 1995 and 1997. The quoted passage is one (the longest) of three published on the Internet by blogger known as Lady Pasztet.[22] She stumbled upon those passages when collecting material for her university thesis and routinely browsing literature on Africa. Two manifest types of positions could be distinguished in the controversy: the dominant moral position and the less prominent scientific position. In the first emotionally charged opinions prevailed, condemning or defending the journalist. The second largely included the general question about conditions of admissibility of such utterances like those quoted by Lady Pasztet. Opinions of the second type revolved around the paradox of the text's racist spirit coupled with distinctively cosmopolitan, non-racist attitude of the author, known as a music journalist. The first type of

---

21 Marcin Kydryński, *Chwila przed zmierzchem* (Warszawa: Prószyński i S-ka, 1995), pp. 109–110.
22 The original post apparently has been removed, but its extensive copied fragment can still be found at the following address (9 Sep 2021): http://jadrociemnosci.blogspot.com/2010/11/komentarzowka-11.html?view=snapshot&m=1.

contributions used predominantly individualistic ontology and ethics (personal responsibility, predilections etc.), the second rather holistic ontology drawing from social sciences (cultural anthropology, sociology), stressing systemic agency and wide cultural phenomena (like Eurocentric views).

As far as it is possible, I would like to skip the first context, which is moral discussion, altogether and focus on the second. More precisely, I would like to refer to the moral aspect of the controversy at most as a subject of analysis, that is, viewing it from the meta-level as a part of the cultural situation that is to be analysed. However, to explain to what degree theory had fostered the Holmes' effect, which is seeing something else, for the sake of reasoning I must descend from the meta-level to the object level, but within the discussion of the second kind (scientific approach). Nevertheless, I would like to stress that my goal strictly remains an answer to the question "What does theory do?" by using a concrete case of socio-scientific controversy, putting aside the controversy itself.

For the purpose of showing theory's role as allowing to see something else, I purposefully quoted *Orientalism* first and next *Chwila przed zmierzchem*, additionally using emphasising fragments in both. The phenomenon described by Said in his book is clearly noticeable in *Chwila przed zmierzchem*. In other words, the latter is an example of a "Orientalist" attitude and communication.

To fully see this let us juxtapose the emphasised fragments of both texts:

Said's *Orientalism*	*Chwila przed zmierzchem* (an example of an orientalising text)
"This is especially evident in the writing of travelers and novelists: women are usually the *creatures of a male power-fantasy*."	"I have always thought that there is a certain regularity in *white men's desire to possess black women*."
"They *express unlimited sensuality*"	"*And they are unimaginably beautiful.*"
"they are more or less *stupid*"	"Their girls have distrust inscribed in their genes. They are skittish like small antelopes, while they take their beauty after big cats. They are not suitable for domestication. They speak a different language even when they use the same words as us."

	"They had run away from their tribes to big cities or had already been born there and *they want to live to the rhythm of their pulsating bodies.*"
"above all they are *willing*"	"Most often, though, they do not speak in the presence of men. They nuzzle up to them. Or they hunt, like lionesses. African prostitution is this kind of hunting. For most Africans sex is life's only pleasure. They use it in a way just as natural and spontaneous as animals. If they do not work and do not make love, boredom kills them."
	"Their prostitution is not consequent. They take money when they need it. Sometimes they only ask for breakfast. *They frequently simply ask to be possessed.*"
	"For it is not human charm, to which we are accustomed, but animal charm. These girls ask to be covered. They begin then to smell differently, suddenly, like halved fruit. *Everything in them speaks about intercourse, eyes, lips, hands.* At last they talk as cat females would speak if they could."

This is simple illustration, and let us remember that I am recalling only very small part of the discussion. If we were to examine the text more thoroughly, sentence by sentence, we would end up with much more material.

The timeline in which the controversy is embedded is also significant. Said wrote his book mainly in the USA, in mid-1970s, and it was published in 1978. Its first Polish translation appeared in 1991 in PIW's (Polish Publishing Institute's) well established and respected "+/- ∞" series. The second translation was published in 2005. *Chwila przed zmierzchem*, as I have already mentioned, got published in 1995 and its second edition in 1997. The controversy sparked in 2011. What does this timeline tell us?

We can determine the timeframe and reconstruct key chains of events. It shows for example the fact often overlooked, that theories travel, and to travel they need vehicles. And ANT is quick to teach us, that travels of this kind change a bit of what travels. As we see, it took thirteen years for Orientalism as a theory to cover the distance from the primary American academic milieu to Polish bookshops, or maybe even more, if we assume the initial date to be the beginning of Said's work, not the date of the first English-language edition. However the very physical act of making a journey does not have to mean much. Let us note that while by early 1990s *Orientalism* had exerted significant influence on postcolonial studies in the world, it did not interact in any form whatsoever with *Chwila przed zmierzchem*, published four years later than its Polish edition. At least we know nothing about such an interaction. In the course of the 2011 dispute some participants remarked that after its publication the book had not provoked the kind of reactions it did several years later.[23]

The lesson we can learn from it is this: readers from 1995 and 1997, or more generally, from 1990s could not see what was noticed by Lady Pasztet (who's our Sherlock Holmes in this case) and later other readers. The same thing that made impossible even for opponents – like the *Sherlock* series' Philip Anderson character – to ignore the new reading of the book. Now they were not able to unnotice this "new" contents, all they could really do was switch to a polemic mode.

This is the way theory worked. Both Lady Pasztet and numerous researchers around the world before, having read Said's book, began to notice something else in the reality surrounding them, in their everyday lifeworlds.[24] We can only suppose that in some cases Said gave a name for what they could already see and experience, but they did not have appropriate notions at their disposal to capture it. In other cases he constructed with the help of the technics of his discipline, and along with other researchers in postcolonial studies, completely new elements of lifeworlds, like in the example above for lifeworlds of cultural anthropology students.

---

23 The author speaks about it in his statement from 16.02.2011: "When the book was published, it was reviewed, described, read on the radio's Programme 3, and no one took offence in the passages which now, stripped of context, spark such interest," (15 Mar 2016) http://www.kydrynski.com/oswiadczenie.html, currently removed.
24 I refer here to the phenomenological concept as developed by Peter Burger and Thomas Luckmann in their *The Social Construction of Reality* (Peter Berger and Thomas Luckmann, *The Social Construction of Reality: A Treatise in the Sociology of Knowledge* (Garden City NY: Anchor Books, 1966)).

3. Action 2: Forget About Your Work

This is not all, though. We could rightly add that the problem of the relation between seeing and knowledge has a long history in philosophy of science, as prove Ludwik Fleck's classic analyses.[25] Moreover, at the first glance the tension between readymade science and science in action is not visible here. Therefore to become fully aware how theory allows to see something else, we must pay attention to yet another element in Lady Pasztet's Holmes-like flash of recognition.

It is about scanty commentary. Like a good anthropological observer, she lets the investigated reality speak rather than claiming that privilege to herself, when she says:

> Looking for material for my master's thesis, I routinely skimmed Polish writers' books about Africa. From book to book, from note to note, I stumbled upon Marcin Kydryński's book *Chwila przed zmierzchem*. It is an account of his journey to Africa plus a handful of the narrator's (hero's) reflections. Basically these passages do not require commentary. Reading this book is accompanied by disbelief. This is Marcin Kydryński's beautiful, deep voice, one of an outstanding connoisseur of Warsaw's salons and jazz, Anna Maria Jopek's husband, his children's father, an ambitious photographer with a cross on his chest (jutting out of his tightly fitted shirt)… and such phrases! A good review should encourage reading, however it is not suitable to quote it. I hope I will not ruin anyone's reading surprise with these three tiny fragments.

In the end she adds:

> The book had two editions: in 1995 and 1997. It provoked virtually no discussion. It is 2010 now, controversies about postcolonialism, orientalism or racism arise everywhere – maybe we should give the book one more chance and discuss Marcin Kydryński's work.

The whole theoretical content boils down to a meta-remark from the third sentence of the latter fragment: "controversies about postcolonialism, orientalism or racism arise everywhere." She does not say: "Here I recognise orientalism / racism / postcolonialism because" Instead, she only leaves a suggestion, directs reader's attention. And before that she adds: "Basically these passages do not require commentary."

---

25 Ludwik Fleck, *Psychosocjologia poznania naukowego. Poznanie i rozwój faktu naukowego oraz inne pisma z filozofii poznania* (Lublin: Wydawnictwo Uniwersytetu Marii Curie-Skłodowskiej, 2006), pp. 214–231 and 306–323. See also the English translation of his 1935 book, which is main part of the quotet above *Psychosocjologia poznania naukowego*: Ludwik Fleck, *Genesis and Development of a Scientific Fact* (Chicago and London: The University of Chicago Press, 1979).

This disappearance of theory and its reduction to three catchwords ("postcolonialism, orientalism, racism") does not testify to the lack of theory, and presence of hard, down-to-earth reality, practice opposing theory, brutal racism as opposed to academic work detached from life. On the contrary, it is that scarce presence that hints at theory's strong influence (postcolonial studies, orientalism etc.). Since it is only pointed at, as if it was something in the air, as if it was omnipresent ("controversies … arise everywhere"), so that everyone catches the drift immediately, when the slight gesture is made.

Here I would see the second type of theory's workings: its disappearing. When the job is already done and people can see something else in the everyday world they inhabit, the weight of analyses, arguments, and examples is already unnecessary. The simple, quick reflex of recognition is all that is left. And this is how I see the role of theory. Not theory itself, obviously, but theory as a force influencing an individual, influencing what in epistemology is traditionally called a subject.

Here we reach the moment that lets us understand why the ultimate goal of theoretical work does not have to be a sort of a construction in the vein what S. Nowak outlined. Certainly, it is possible that certain groups of readers would most efficiently pass from digesting theory to its disappearance with the aid of this particular form (i.e. stable, coherent structure of propositions). However, as shown by the example of Said and many other that come to mind, theoretical work meant to be completed with its own disappearance may operate in many different ways: show captivating examples, paralyse with steel-hard logic, convince with references to literature, compel to scrupulous text analyses etc. A theoretical reasoning that fails to disappear from view, will still burden the memories and minds of its bearers. In other words, where disappearance fails, positive modalisation (in ANT sense) of primary statements will not be completed, facts will not be fabricated either.[26]

There is a certain important difference between a skilled researcher using theory and its merely competent user, presented in the general context of acquiring competences, in Bent Flyvbjerg's *Making Social Science Matter*,[27] as a difference between someone who has mastered a craft and someone who is still a student. The master, thanks to knowledge and experience, captures a particular

---

26 On the issue of modalisation in scientific texts and building facts I refer again to: Abriszewski, *Wszystko otwarte*, pp. 27–42, Latour, *Science In Action*, pp. 21–62, Latour and Bastide, "Writing Science," and Latour and Woolgar, *Laboratory Life*, pp. 151–186.

27 Bent Flyvbjerg, *Making Social Science Matter. Why social inquiry fails and how it can succeed again* (Cambridge, New York: Cambridge University Press, 2001).

situation "intuitively," that is she or he notices many elements with a "single" glance. An advanced learner with some experience is also able to identify certain phenomena thanks to making adequate effort. However, and here is the profound difference, the master can subject his or her insights to analytical verification. This means that the apprentice analytically dismantles the situation and the master can do it with his or her own conjectures. I suspect that Lady Pasztet, as a competent user of the theory, correctly and quickly recognises the investigated phenomenon, but, contrary to someone who masters a theory, does not automatically launch the analytical, "auto-check" mode. That is why the post begins with descriptions of two situations: searching for material and the experience of listening to the radio. This is the evidence for the thesis that an important moment in theory's workings is its disappearance.

4. Action 3: Circulation

You might have already noticed that in the course of argument, the place of the word "theory" slowly gave way to "theoretical work." Maybe this kind of shift more precisely reflects the difference introduced in the beginning between ready-made science, described by Nowak with its closed edifice of theory, and science in action with an array of actions that I wish to present here.

It is similarly difficult to capture the next moment of theoretical work or indicate what theory does apart for letting us see something else and then disappearing. Thus we have seen so far, on the one hand a dose of communication surplus was necessary for subsequent people to notice something else, on the other hand, though, a possibility to proceed in the opposite direction: reduce the whole theoretical construct to one or two notions, so that not just the surplus would disappear, but also everything that we traditionally considered to be theory. The third moment must at the same time connect these divergent elements, and transfer them to one's own everyday lifeworld.

In other words, a theory must also circulate.[28] Its ability to make people notice something new will be of no use if it stays in one place (e.g. on a bookshelf in a university library). Certainly, it is not possible to attribute causal power in this respect, it depends on the medium in which it was coded, communication structures, relations of power and authority, and many different factors. But the form it is expressed in is its integral part as well, other than in Nowak's

---

28 Latour's *Pandora's Hope* is obviously a contemporary classic devoted to circulation of theories in various environments. Bruno Latour, *Pandora's Hope: Essays on the Reality of Science Studies* (Cambridge, London: Harvard University Press, 1999).

understanding, where rigidity and uniformity of form (propositions, theorems, laws) was supposed from the methodological perspective to see only the theory's content. If we use, though, the STS/process switch I recommended and follow live science in action, we will see distinctly that the rigidity is a result of enforcing a specific framework to the object of research from the outside by a methodologist or a philosopher, it is a tool, not part of the investigated object. If we heed the ANT slogan "follow the actors," we will also follow theory (or theoretical work), then not only it will not get locked within such a framework (which we have seen before), but also it will turn out that operating codes and media is an integral part of theoretical work, too.

The same process, the same function of theory can also be expressed in other, more traditional, categories, referring, by the way, to the individual/community opposition. According to the approach whose example was Nowak's textbook, theory is something impersonal. The subject – creating it, interpreting it, applying it, reshaping it – basically disappears from view. The question "what does theory do" brings the subject back to the scope of investigation. I presented the two first kinds of actions (see something else and disappear) from the point of view of an individual researcher. But just as science is a collective effort, theory too must encompass a broader range of entities than only one single theoretician. Therefore "circulation" is meant to add a community dimension to the first forms of theory's activity.

Here I use the old opposition between individual and community only for the sake of investigating theoretical work, because I do not think ontologising this approach is well justified. Thus I do not refer to particular activities of theory using an "individual" and a "community" as labels, but speak of circulation, assuming that what we call individual and community emerges as results of the circulation and constructing new relations.

A certain duality is characteristic of circulation of a theory: on the one hand theory (theoretical propositions, utterances etc.) may have the power to enlist subsequent individuals – those won for the cause, who will begin to see something else, but not only them. On the other hand, in compliance with the rule saying that each translation is transformation, a theory can be subject to change and mutate. I am inclined to treat it as its inherent feature. What is more, in case of humanities I would assume that a typical situation is emergence of such a theory which, to paraphrase Annemarie Mol, is more than one, but less than many.[29]

---

[29] Annemarie Mol, *The Body Multiple: Ontology in Medical Practice* (Durham, London: Duke University Press, 2002).

Differences inscribed in the thought collective operating with a particular theory would not then be an obstacle, noise, or an error, but a normal element of a theory's workings. Furthermore, what is unusual, is full unanimity. It is relatively easy to trace the theoretical dispersion in our example: if we have a closer look at different readings of *Chwila przed zmierzchem*, we will see that most readers adopted the perspective of postcolonial studies, but they did it in different ways and arrived at various conclusions and questions. And this is that specific moment in the theory's circulation.

A theory may use different mechanisms enabling its circulation, or, in ANT terms, enlisting others. Theories might simply be directed at readers-scholars interested in a given area, like for example various epistemological theories are likely to attract philosophers-epistemologists. Another group of theories in order to stimulate circulation might aim at focussing around an issue interesting for particular categories of the public, like theoretical scenarios analysing demography in Poland in the few next decades, or theories describing mechanisms of emotional agitation in addressees of advertisements. Still other theories may take advantage of controversiality of their results or their objects of study, like for example non-biological explanations of ADHD. There will also exist theoretical works whose attracting power will flow from the ability to redefine important areas of human experience, and here I would include Said's *Orientalism* and postcolonial theory in general. Each of these forms can have significant influence on the way the theory is formulated, who is its virtual addressee (for example who is an opponent, who is to be convinced), and so can the choice where to publish the findings (through the profile of the magazine, department, publishing house, editorial series).

5. Action 4: Engage Emotions

There is, however, a certain mechanism that plays an important role both in supporting a theory's circulation (enlisting others) and in the two formerly described actions: in enabling to see something else and in disappearing. Traditional discourses on science leave the question of emotions outside their interest area, they only allow it in the context of discovery (e.g. the discoverer's emotional tension during research) and in the domain of a theory's aesthetic value (beauty and elegance of optimally constructed theories). I think, though, that there is much more to that in theoretical work, and that emotions constitute a common and vital component of what theory does. Let us read once again one of previously quoted passages:

> Basically these passages do not require commentary. Reading this book is accompanied by disbelief. This is Marcin Kydryński's beautiful, deep voice, belonging to an outstanding connoisseur of Warsaw's salons and jazz, Anna Maria Jopek's husband, his children's father, an ambitious photographer with a cross on his chest (jutting out of his tightly fitted shirt)... and such phrases!

The whole passage above revolves around emotions. There is an explicit expression of "disbelief" in the second sentence, next comes an attempt at engendering the emotional turmoil provoked by *Chwila przed zmierzchem* in the reader, by contrasting different contexts which the book author spans. Moreover, the first of the sentences ("these passages do not require commentary") conveys a similar message by apparent speaking about something else. For if any comment is unnecessary, than why saying that? So the message is performative, it is located not in the contents of the words but in the very fact of saying (writing) them. The said message seems clear and "obvious," and it is meant to inform that "they are despicable and scandalous." By the way, direct articulations of such feelings could be found in numerous posts, where debates went on about interpretation of *Chwila przed zmierzchem*, especially in the discussion mode that I have named moral, which, of course, is not surprising, because axiological debates quickly activate the emotionally charged mode.

It is not difficult to make a connection between theory and emotions, however provided that we abandon the old reason/heart opposition, and employ a more contemporary approach, based on cognitive science. For this purpose I would like to utilize the approach developed by Noël Carroll, a philosopher interested in popular culture.[30] A significant part of his book *A Philosophy of Mass Art* is devoted to a critical discussion of the issue of emotional involvement, which leads him to abandoning the so-called concept of the identification with the hero of a pop-cultural story. Instead he proposes his own concept, developed on the foundation of cognitive science, which he calls the cognitive theory of emotions.[31] His concept is tersely summarised in the following paragraph:

> emotions are intimately related to attention. ... I have suggested, furthermore, that the emotions are related to our attention-focusing mechanisms. They direct our attention to certain details in an array, rather than to others. The emotions enable us to organize those details in significant wholes or gestalts, so that, for example, our attentions selects

---

30 Carroll himself prefers the concept of "mass art." I prefer, though "popular culture" instead, as currently the most widespread and avoiding reference to historic phenomenon of masses. Nevertheless, this is not the place for the thorough discussion of Carroll's position, and let's postpone it for some better occasion.
31 Noël Carroll, *A Philosophy of Mass Art* (Oxford: Clarendon Press, 1998), p. 250.

up or battens upon the concatenations of details in the situation that are, for instance, relevant to harm or to misfortune. The emotions operate like a searchlight, foregrounding those details in a special phenomenological glow. And, as well, once we are in the midst of an emotional state, we not only hold to those details, often obsessively, but we are also prompted to search more details with similar relevance to our presiding emotional assessment of the situation. The emotions manage our attention when we are in their grip. And the management undergoes changes in the sense that it first alerts our attention to certain gestalts and hold our attention on them and then encourages further elaboration of our attention, inclining us to search for further element of the relevant gestalt in the stimulus, and leading us to form expectations about the kinds of things we should be on the lookout for as the situation evolves.[32]

In this approach it is impossible to oppose emotional reception of a particular message to rational reception. Instead, what is crucial here is the emergence of a certain link connecting the environment, situation or some "external" object and the organism's (individual's, recipient's etc.) internal state. Without emotional agitation this link will not establish (I put away a book which I am not interested in, I turn off a boring film etc.), or it will be very weak. The element mediating between the internal state and an external object are our cognitive states.[33] This means that

> an emotion is made up of at least two components: a cognitive component such as a belief or a thought about some person, place, or thing, real or imagined; and a feeling component (a bodily change and/or a phenomenological experience), where, additionally, the feeling state has been caused by the relevant cognitive state, such as a belief or a belief-like state.[34]

Thus, it means that any artistic narration: a film, a story, a graphic novel, or a theatrical play, but also a theoretical narration can play a role of such a cognitive component within our everyday experience. In other words, this framework linking the cognitive component with emotional states lets us see the analogy between the reader's indignation in response to the main character's lot in one of Clint Eastwood's westerns, where a group of people looking for stolen cattle hangs him without a trial (to no effect, as it soon turns out), and a similar indignation in a reader of *Orientalism* who learns details of, say, racism in some European art.

With this in mind, the concept of Orientalism simply offers a complex, although possible to simplify (as we have seen in the section about the

---

32 Carroll, *A Philosophy of Mass Art*, pp. 261–262.
33 Carroll, *A Philosophy of Mass Art*, p. 253.
34 Carroll, *A Philosophy of Mass Art*, p. 254.

disappearing) cognitive component which teaches to recognise other cases of a structure engendering the reader's rage, sense of injustice, indignation, disgust, or the contrary, the satisfaction that justice has finally triumphed etc. As Carroll says, "what emotional state I am in is determined by my cognitive state,"[35] and further: "Having the relevant cognitive states is a necessary condition for being in these emotional states."[36] It means that in order to trigger a specific emotional state, a cognitive state that is crucial for that must appear first. Moreover, the latter can also modify emotions when it changes itself, e.g. when indignation gives way to satisfaction with compensation for the harm and punishing iniquity. I claim that this mechanism works also for scientific theories, and in humanities and social sciences it may even be their important component. It is not difficult to see it in the case of the theory of Orientalism and postcolonial studies founded on this very structure of justice infringement and a certain primary harm that must be recognised, labelled and atoned for.

The first quote from *A Philosophy of Mass Art* says something else, though: along with linking cognitive and emotional elements, emotions actively redirects our attention. Thanks to this observation we can, at least hypothetically, explain how the controversy on *Chwila przed zmierzchem* started. Having read Said's book not only implied being imparted of a specific portion of knowledge, but this knowledge was crucial, as it interplayed with a specific emotional state, for example indignation with colonial injustice. The moment that I have previously called "disappearance of a theory" entailed creating that "whole or gestalt" that Carroll spoke about. From that moment on, the emotion was ready to "act" whenever the cognitive state linked to it appeared, that is when the reader comes across something resembling situations analysed by Said. Consequently, this preparation made the reader's attention more acute and directed her at "suspicious" cases. *Chwila przed zmierzchem* turned out to be such a case.

If my supposition as to the link between theory in humanities and social sciences and emotions is right, we can see that it encompasses quite a substantial area of academic disciplines, to name a few: postcolonial studies, feminist studies, critical theories of all sorts, queer studies, animal studies, revisionist approaches in history (for instance using class theory), all that is occasionally described today as "new humanities," a sizeable chunk of critical tradition

---

35 Carroll, *A Philosophy of Mass Art*, pp. 254–255.
36 Carroll, *A Philosophy of Mass Art*, p. 255.

and probably many other trends. What repeats in all of them is the structure of restoring justice, so typical to many products of popular culture.[37]

I emphasize here, like in earlier stages of my argument, that the emotional aspect of theoretical work in humanities and social sciences should be should be viewed as something normal, as their ordinary component and not a pathological surplus, or redundant noise in the message, which should be removed or ruled out. Following Carroll's perspicacious observations a step further, one can extend another feature of popular culture narrations to theoretical work, something that Carroll calls "criterial prefocusing:"

> Whether verbal, visual, or aural, the text will be prefocused. Certain features of situations and characters will be made salient through description or depiction. These features will be such that they will be subsumable under the categories or concepts that, as I argued earlier, govern or determine the identity of the emotional states we are in. Let us refer to this attribute of texts by saying that the texts are *criterially prefocused*.[38]

Let me say it once again: the way of conducting a theoretical work aims at steering the reader's attention. This is prefocusing, directing the addressee's attention, with means embedded in the message. The reader encounters subsequent elements, which through their specific cognitive content engender trigger emotional states. The contents of the theoretical texts is certainly different than the contents of action films or science-fiction novels, but, as we have seen in the case of Orientalism, the mechanism remains the same. We can figure out what the theory of Orientalism has brought to the reception of quoted fragments from *Chwila przed zmierzchem*. In "naive" reading one could emotionally understand the narrator's fascinations, interpreting them as a poetic expression of an Eastern European traveller's experience of Africa. For some readers in the 1990s it could even provide an *ersatz* of "European subject" ("being a true European") by impersonating an imagined subject position in the fantasy of white man from Europe in black Africa. In turn, in the reading "equipped" with the theory of Orientalism, the reader feels anger, indignation, distaste, when approaching the same fragments, which further directs her attention towards all symptoms of

---

37 Carroll writes: "Anger, for example, appears across mass-art genres and is remarkably pervasive." and further on the same page: "Hatred, one is tempted to say, is a major calling-card of mass art. Furthermore, the anger that we bring to bear on antagonists – because of the wrongs they inflict on the protagonists – also encourages a taste of revenge in us, inviting us to stay with the story in the hope that the villains will be dealt their well-deserved punishment." (Carroll, *A Philosophy of Mass Art*, p. 281).
38 Carroll, *A Philosophy of Mass Art*, pp. 263–264.

orientalising: the animal metaphors, construction of the exotic Other, the relations of power embedded in the narrator's perspective, Eurocentrism.

We can see then that the widespread belief that theoretical arguments belong only to the realm of rationality, radically separated from the realm of emotions, or only marginally getting in touch with it, is simply wrong. Theoretical works presuppose (consciously or not) a specific structure of emotional reactions to the argument. As it is with any message, the real addressee may overlap with the ideal one or be missed completely. In the latter case such a theoretician may be called an "ideologist" or a "radical who does not operate with rational arguments." But this will happen only because the opponent will not be won by the emotional structure of the theoretical work, because cognitive components will not engender the "planned" emotional states. Quite the opposite, the result would be then typical academic anger at the "rubbish and gibberish spread by fools and ignorant."

6. Action 5: Reshape the World

Conclusions from the ruminations above on what theory does in humanities will differ depending on how many new, not traditional assumptions reader is going to accept. I skip the case of wholesale rejection, because there is hardly anything to discuss here. First, then, a reader who is not discouraged with the processual switch will accept that studying science in action, including theory in action, can bring interesting results. Although this assumption is likely to be relatively unproblematic (After all what is wrong with examining what scientists do, even if later we are going to give priority back to analysing ready-made science?), matters may take a surprising turn in the next steps. For if we follow the path of STS, especially ANT, and begin to search for theory in humanities, it will turn out that we will need to change our expectations. Only in some cases theory will assume the form of a finished structure of propositions, theorems, laws and definitions. More often than not it will be theoretical arguments, considerations, corrections, generalizations, often if form of fragments rather than full blown, ultimate structure. Shortly, the place of theory is taken here by what I called "theoretical work." Having accepted that, another the question – about "what theoretical work does?" – will become much more important, and the following problems will open: what is our aim in humanities, when we indeed proceed with theoretical work?; what is the purpose of this work if not some elegant, theoretical structure of propositions emerging in the end of the day?; what do we do actually do when our efforts aim at creating and sharing knowledge? Why, how and in what way do we teach our students theory and theoretical practices?

How should we interpret relations between theoretical and empirical practices? How to treat links or transitions between various theoretical schools? And so on.

Secondly, the reader may not only agree to employ the processual switch that I suggested, but also agree with large part of my observations. Yet at the same time said reader may prefer to stay on the epistemological side of humanities, to use the concept developed by Mariola Kuszyk-Bytniewska.[39] He or she will admit that theories in humanities in many ways do construct and reshape our *images* of the world. Such a reader would agree to pay attention not only to structure of theory, but also to processes of theoretical work, staying at the same tome on the side of epistemological constructivism, an approach ubiquitous in humanities and social sciences. Then the name of the game would be world-image, perspective or representation of the world.

Thirdly, it is possible that the reader will fully accept the suggested point of departure, including the *ontological* constructivism of ANT and a significant part of Science and Technology Studies. In terms of Kuszyk-Bytniewska's concept this means entering the area of "social onto-epistemology." With this approach the above four forms of activity which can be associated with theory: (1) see something else, (2) erase elaborate analysis and reduce it to one slogan, (3) circulate and (4) engage emotions, should be completed with the fifth: (5) reshape the world. Therefore, in certain particular cases, operating representations of the world would ultimately affect the order of beings itself. And it is not that the minds of those inhabiting the social world would be changed through new content. The transformation would reach much further: the very content of the world would change. In this sense, Orientalism as a theory (theoretical practice, to be precise) allows to see something new, it can be reduced to a single slogan, it effectively circulates (in its material vehicle, that is Said's book), it engages emotions, but in the end it comes up as a new inhabitant of the collective world. One of its, possibly paradoxical, features, is that on the one hand it will seem independent of any activities undertaken by individuals. It will exist whether or not I decide to teach students about it or write a paper about it. On the other hand, it will obviously be prone to any uses, at the mercy of those who will wish to enable its circulation. There is, though, a third side to it: it will be equipped with a special agency – it will be able to endow people with extraordinary powers that may prove useful in coping for example with literature or films, whose action is placed in the (mythical) Orient. Simultaneously, Orientalism as a social phenomenon will form part of life, as an array of various material practices,

---

39 Kuszyk-Bytniewska, *Działanie wobec rzeczywistości*.

prejudices, convictions, myths, imagery, but also representations recorded on paper, celluloid, canvas or as bits of information. This Orientalism will also be able to create its own forms of agency, enforce specific conduct on people, even contrary to their "deep," "personal" intentions or beliefs, which was repeatedly stressed in debates on *Chwila przed zmierzchem*, indicating at its author's great respect and sympathy for artists coming from Africa.

According to a simple argument, theoretical innovations in natural sciences, at least in some cases, come to our everyday lifeworlds as various objects, artefacts or technologies. One can operate them without any knowledge about theories that lead to their birth. In case of humanities and social science a similar, though slightly different, process occurs. Obviously, full-blown theories with all the subtleties, even the most beautiful concepts in ethics or aesthetics, cannot fully settle in everyday lifeworlds of those people who normally do something else than theoretical work in human sciences. "Artefacts" of humanities will inevitably anchor in the lifeworlds in other ways than their creators have conceived: as quick insights showing something new, as single slogans, as knowledge passed further (or that blocks something else from passing), and as emotion operators. According to this approach theories would participate in births of new actors (or actants), and theoretical texts will be ontological generators programming (like some social "DNA") a new collective world.

Bibliography

Abriszewski, Krzysztof. *Wszystko otwarte na nowo. Teoria Aktora-Sieci i filozofia kultury*. Toruń: Wydawnictwo Naukowe UMK, 2010.

Berger, Peter L. and Luckmann, Thomas. *The Social Construction of Reality: A Treatise in the Sociology of Knowledge*. Garden City NY: Anchor Books, 1966.

Carroll, Noël. *A Philosophy of Mass Art*. Oxford: Clarendon Press, 1998.

Fleck, Ludwik. *Genesis and Development of a Scientific Fact*. Chicago and London: The University of Chicago Press, 1979.

Fleck, Ludwik. *Psychosocjologia poznania naukowego. Poznanie i rozwój faktu naukowego oraz inne pisma z filozofii poznania*. Lublin: Wydawnictwo Uniwersytetu Marii Curie-Skłodowskiej, 2006.

Flyvbjerg, Bent. *Making Social Science Matter. Why social inquiry fails and how it can succeed again*. Cambridge, New York: Cambridge University Press, 2001.

Gandhi, Leela. *Postcolonial Theory. A critical introduction*. Crows Nest: Allen & Unwin, 1998.

Grobler, Adam. *Prawda i racjonalność naukowa*. Kraków: inter esse, 1993.

Grobler, Adam. *Metodologia nauk*. Kraków: Wydawnictwo Aureus, Wydawnictwo Znak, 2006.

Kmita, Jerzy. *Essays on the Theory of Scientific Cognition*. Dordrecht: Springer-Science+Media, 1991.

Kuszyk-Bytniewska, Mariola. *Działanie wobec rzeczywistości. Projekt onto-epistemologii społecznej*. Lublin: Wydawnictwo Uniwersytetu Marii Curie-Skłodowskiej, 2015.

Kydryński, Marcin. *Chwila przed zmierzchem*. Warszawa: Prószyński i S-ka, 1995.

Latour, Bruno. *Science in Action. How to Follow Scientists and Engineers through Society*. Cambridge Mass.: Harvard University Press, 1987.

Latour, Bruno. *Pandora's Hope. Essays on the Reality of Science Studies*. Cambridge, London: Harvard University Press, 1999.

Latour, Bruno and Bastide, Françoise. "Writing Science – Fact and Fiction: The Analysis of the Process of Reality Construction Through the Application of Socio-Semiotic Methods to Scientific Texts." In: *Mapping the Dynamics of Science and Technology*, eds.: Michel Callon, John Law, Arie Rip. London: The MacMillan Press Ltd., 1986, pp. 51–66.

Latour, Bruno and Woolgar, Steve. *Laboratory Life: the Social Construction of Scientific Facts*. Los Angeles, London: Sage, 1979.

Law, John. "The Heterogeneity of Texts." In: *Mapping the Dynamics of Science and Technology*, eds.: Michel Callon, John Law, Arie Rip. London: The MacMillan Press Ltd., 1986, pp. 67–83.

Loomba, Ania. *Colonialism/Postcolonialism*. London and New York: Routlege, 2015.

Mol, Annemarie. *The Body Multiple: Ontology in Medical Practice*. Durham, London: Duke University Press, 2002.

Nowak, Stefan. *Methodology of Sociological Research. General Problems*. Dordercht/Boston, D. Reidel Publishing Company Warszawa: PWN – Polish Scientific Publishers, 1977.

Nycz, Ryszard. *Kultura jako czasownik. Sondowanie nowej humanistyki*. Warszawa: Instytut Badań Literackich, 2017.

Sady, Wojciech. *Spór o racjonalność naukową. Od Poincarégo do Laudana*. Wrocław: FUNNA Sp. Z o.o. – Wydawnictwo, 2000.

Said, Edward W. *Orientalism*. London: Penguin Books, 2003.

Suppe, Frederick. *The Semantic Conception of Theories and Scientific Realism*. Chicago: University of Illinois Press, 1989.

Taylor, Charles. *Philosophy and the Human Sciences. Philosophical Papers 2*. Cambridge: Cambridge University Press, 1985.

Aleksandra Derra

Women Not to Be Seen: The Medical Versus Feminist Thought Style on Anorexia Nervosa[1]

Abstract: The authoress uses Ludwig Fleck's analytical tools from his psychosociology of scientific knowledge (such as a thought style, thought collective) to present the fundamental processes that have accompanied the development of the phenomenon of anorexia nervosa. Notably, she follows Flecks's findings of the stabilization of scientific facts and applies them to the history of anorexia and the early attempts at naming and diagnosing it. She confronts anorexia as a disease in a medical thought style (an individual, homogenous disorder of psychological etiology) with anorexia in a feminist thought style, as an illness where social and cultural factors play a significant role. Given that about 90 % of anorexia patients are women, feminists underline that it should be treated as an extreme version of a more common affliction of our culture (with its obsession with thinness, good looks, attractiveness). They notice that it cannot be cured on the individual level, for more systematic changes in thinking about subjectivity, femininity and corporeality are required. It seems that there are no effective methods of healing anorexia in the medical thought style; in contrast, the feminist view raises hopes for postulating new ways of treating this phenomenon, which gives us possibilities of new research and new ideas for eliminating it.

Keywords: anorexia, feminism, thought style, thought collective, medical thought style, feminist thought style, disease, corporeality

In science, just as in art and in life,
only that which is true to
culture is true to nature

– Ludwig Fleck[2]

In what follows, I use analytical tools and basic concepts of Ludwik Fleck's psychosociology of scientific cognition (such as thought style, thought collective, scientific fact) to show the key processes accompanying the occurrence of

---

1 I published my research on this topic primarily in the article written in Polish entitled "Kobiety jako czynnik przemilczany. Wybrane wątki epistemologiczne powstawania medycznego zjawiska anoreksji," *Przegląd Filozoficzny* 2 (74)/2010, pp. 27–44. Here its mayor part were translated into English, and it has been amended, redrafted and updated accordingly.
2 Ludwik Fleck, *Psychosocjologia poznania naukowego. Powstanie i rozwój faktu naukowego oraz inne pisma z filozofii poznania* (Lublin: UMCS, 2006), p. 65.

the medical phenomenon of anorexia.[3] To this end, I briefly present the history of this phenomenon known from the literature on the subject, early attempts at naming, specifying, and diagnosing it, stressing the role of the dissemination and ongoing validation of a medical thought style. The classification of anorexia as a disease (an individual, homogeneous mental disorder) in a medical thought style will be confronted with a feminist thought style, in which the role of social and cultural factors in its etiology is emphasized while noting that 90–95 % of patients are women. Here we talk more about the illness being experienced by a person situated in a complex and complicated social and cultural surroundings. This style treats anorexia as an extreme version of a universal affliction of our culture, which cannot be eliminated at the level of the individual without systemic changes in the way of thinking about subjectivity, femininity, and carnality (the question of whether such changes can be achieved at all is a separate matter). Anorexia as a disease embedded in medical discourse has become a common fact, a fact that is of little use to epistemological research.[4] Since, as it seems, there are no effective methods of curing anorexia based on medical findings, feminist thought style gives hope for the opening up of new ways of thinking about anorexia, thus offering the possibility of conducting new studies on it, including developing alternative methods of its elimination.[5]

In the light of the approach proposed by Fleck, anorexia nervosa can be termed a scientific fact as it represents a conceptual structure corresponding to a medical thought style,[6] an occurrence of historical-thought relationships created as a result of a specific, in this case medical, thought style.[7] The medical thought style should be understood here as a specialist field of research and an extensive network of practices related to specific institutions and their representatives. These representatives are experts whose actions extend with their influence beyond the narrow circle of specialists and beyond the area of competence

---

3   Originally Fleck published his main and the most famous book in German in 1935 under the name *Entstehung und Entwicklung einer wissenschaftlichen Tatsache. Einführung in die Lehre vom Denkstil und Denkkollektiv*. The first translation in English appeared in 1979 as *Genesis and Development of a Scientific Fact* (Chicago: University of Chicago Press, 1979). Since Fleck used two languages in his scientific work, German and Polish, I refer to Polish versions of his book and later papers.
4   Fleck, *Psychosocjologia poznania naukowego*, pp. 31, 109.
5   Ludwik Fleck, *Style myślowe i fakty. Artykuły i świadectwa* (Warszawa: IFiS PAN, 2007), p. 118.
6   Fleck, *Psychosocjologia poznania naukowego*, p. 107.
7   Fleck, *Psychosocjologia poznania naukowego*, p. 117.

related strictly to medicine (it is enough to mention the legal regulations based on medical assessments). In this sense, anorexia is also a fact of culture.[8] A phenomenon that has become an integral part of mass culture, most of us possess selective but strongly standardized knowledge. That knowledge is most often composed of the belief that anorexia is a disease that consists of refusing to eat and mainly affects young girls, with the primary cause being some "mental instability" that may stem from various sources. Anorexia, as a characteristic feature of our culture, has become an object of research outside of medicine, extending to anthropology, philosophy of culture, history of medicine, not to mention psychiatry, psychoanalysis, or psychology, whose representatives, given the etiology of anorexia established by medical science, have become its researchers and experts. The noticeable presence of the discussed phenomenon in culture leads to treating it as a litmus paper of the psychophysical and spiritual condition of contemporary (mainly young) women, making it possible to conduct much broader reflections on the status of carnality (the myth of appearance, the cult of health, hygiene, and beauty, the social and ethical consequences of the use of state-of-the-art biomedical technologies), and sexuality in contemporary culture.[9] Anorexia, included in the broadly understood spectrum of mental disorders, arouses extraordinary interest in the popular mass media and tabloids (similarly to depression, and contrary to schizophrenia or cyclophrenia) because its effects are not only easily discernible, but it also appears that their presence is a measure of a person's sensitivity and an expression of their character. On the one site, anorexia as a medical phenomenon is perpetuated by invoking the ("discovered" by medicine and once and for all established) causes of its occurrence, characterizing its typical course, and administering the most effective methods of treatment: all of this resulted for a very long time in the popularization of the belief that it is a single-dimensional and etiologically homogenous phenomenon. The inclusion of socio-cultural research on anorexia in the mainstream of defining this phenomenon recognized its complexity, lack of homogeneity and possible spiritual character.[10] On the other side, presenting it as a cultural

---

8   One should note that the scientific fact is, by all means, cultural for Fleck. It is possible to some extent to recreate its history, to show what events led to its constitution, what psychological points of view made it possible to legitimize it in such a thought style and not another. See Fleck, *Psychosocjologia poznania naukowego*, p. 107.
9   Helen Malson, *The Thin Woman: Feminism, Post-structuralism and the Social Psychology of Anorexia Nervosa* (London: Routledge, 1998), p. 6.
10 Pamela I. Swain et al., ed. *Anorexia nervosa and bulimia nervosa: new research* (New York: Nova Science Publishers, 2006); Lindsey Hall, Monika Ostroff, *Anorexia*

phenomenon (commonly known, discussed, exposed, and explored in the media or on the Internet),[11] treating it as an element of the lifestyle. It makes it de-tabooed and stripped of its strictly personal dimension, which is characteristic of diseases. It is dangerous as it effectively hides the fact that anorexia often leads to death (also suicidal), is a life-time-long experience, and independently from healing results, leads to somatic complications resulting from malnutrition.[12]

Being Inspired by Fleck's Approach

When looking at selected feminist and women-oriented studies of the phenomenon of anorexia, one can see that although they are conducted from different angles, they have several elements in common.[13] First of all, the perspective

---

nervosa: A Guide to Recovery (Carlsbad: Gurze books, 1999); Becky Thompson, *A Hunger So Wide And So Deep: A Multiracial View of Women's Eating Problems* (Minneapolis: University of Minnesota Press, 1994); Piotr Maroń, "Standarization of Medicine – case of anorexia nervosa," in: *Graszewicz.com. Media. Komunikacja. Kultura*, ed. Dominik Lewiński, Karina Stasiuk-Krajewska, Roman Wróblewski (Wrocław: Libron, 2017), pp. 211–229; Emma White, *The Spirituality of Anorexia: A Goddess Feminist Thealogy* (London: Routledge, 2019).

11 A good example is the Internet pro-ana movement (abbreviation for "pro-anorexia"), advocating in favor of anorexia, which bases its ideology on Christopher Marlowe's saying "quod me nutrit me destruit" (what feeds me, destroys me), promoting the idea of a very slim figure and preaching hatred for eating. See: Stephanie Tierney, "The dangers and draw of online communication: Pro-anorexia websites and their implications for users, practitioners, and researchers," *Eating Disorders* 14 (3)/2006, pp. 181–190; Grace Overbeke, "Pro-Anorexia Websites: Content, Impact, and Explanations of Popularity," *The Wesleyan Journal of Psychology* 3/2008, pp. 49–62.

12 Debra Franko, Aparna Keshaviah, Kamryn Eddy, Meera Krishna, Martha Davis, Pamela Keel, David Herzog, "A Longitudinal Investigation of Mortality in Anorexia Nervosa and Bulimia Nervosa," The American Journal of Psychiatry 2013. https://doi.org/10.1176/appi.ajp.2013.12070868 (20 Dec. 2021); C. Laird Birmingham, Jenny Su, Julia Hlynsky A., Elliot Goldner M., Min Gao, "The mortality rate from anorexia nervosa," *Eating disorders* 38 (2)/2005, pp. 143–146; Hall, Ostroff, *Anorexia nervosa. A Guide to Recovery*; Swain et al., *Anorexia nervosa and bulimia nervosa: new research*.

13 Susie Orbach, *Fat is a Feminist Issue* (London: Hamlyn, 1978); Sheila MacLeod, *The Art of Starvation* (London: Virago, 1981); Kim Chernin, *The Obsession: Reflections on the Tyranny of Slenderness* (New York: Harper and Row Publishers, 1981); Kim Chernin, *The Hungry Self: Daughters and Mothers, Eating and Identity* (Virago Press: London, 1986); Morag MacSween, *A Feminist and Sociological Perspective on Anorexia Nervosa* (London: Routledge, 1993); Julie Hepworth, *The Social Construction of Anorexia Nervosa* (London: Sage Publications, 1999); Thompson, *A Hunger So Wide And So*

of research, i.e., emphasizing the fact that anorexia is an affliction of (mainly young) women, should be adequately explained. It should be stressed that until the eighties of the 20th century, a gender category did not appear in the theories on the formation and development of eating disorders as a factor characteristic for them, perhaps also explaining their etiology.[14] Also shared is the conviction that anorexia can be treated, not as an individual disorder, but as a socio-cultural phenomenon that could only arise in an adequately configured culture in which femininity, appearance, carnality, and food are precisely defined and perpetuated (the way this occurs will be discussed later). In all these studies, the researchers seek to question the adopted conceptualization of anorexia as psychopathology.[15] To this end, they recall the history of anorexia as a phenomenon constructed at a specific time as an object of the medical sciences and show male medical authorities' role and that medical knowledge's prestige was becoming increasingly crucial in culture. It is easy to see that this resembles how Fleck treats the formation of knowledge and the cognitive subject itself: anti-individualistically, collectively, historically, as the process of constitution of ideas. Let me refer at this point to the most critical aspects of his approach.

Both in his best-known work *Genesis and Development of a Scientific Fact* and his minor writings, he repeatedly stressed that thinking is a collective activity and that medicine, like other branches of human knowledge, is socially constructed. The object of knowledge, on the other hand, neutralized and objectified in traditional epistemology (e.g., neo-positivism), does not play a decisive role, as representatives of different thought styles understand it differently.[16] We cannot understand the beliefs formulated in the thought style of the past centuries because we identify ourselves and are mentally supportive of our contemporary style.[17] It does not invalidate the role of historical research and the function of comparative historical research between styles. On the contrary, like syphilis as analysed by Fleck, anorexia has its history, which reveals its entanglement with the style of a given epoch, shows how its characteristics were influenced by the positions, tradition, or factors rooted in the typical mentality (such as upbringing

---

Deep: A Multiracial View of Women's Eating Problems, Susan Bordo, *Unbearable Weight. Feminism, Western Culture and the Body* (University of California Press: London, 2003); Helen Malson and Maree Burns, eds. *Critical Feminist Approaches to Eating Dis/Orders* (London: Routledge, 2009).

14 Bordo, *Unbearable Weight. Feminism, Western Culture and the Body*, p. 47.
15 Hepworth, *The Social Construction of Anorexia Nervosa*, p. 3.
16 Fleck, *Style myślowe i fakty*, p. 218.
17 Fleck, *Style myślowe i fakty*, p. 108.

or the order of cognition) that dominated at a particular time.[18] Again, before the medical nature of anorexia as a mental disorder was established in culture, there existed various concepts of anorexia specific to the time. The equalization of anorexia with a disease caused by mental factors, which is suggested by adding the term "nervosa" (similarly to the equalization of syphilis with a disease caused by a specific bacterium), required a long and complex development of medical thinking (with a significant influence of the accompanying /popular thinking), in which the category of a "disease entity" was developed, ensuring its unambiguous definition. Anorexia is an excellent example of how much cultural effort is needed to dispel doubts about the very concept of disease or conceal the arbitrariness and randomness of the division between the normal and the pathological. Fleck was well aware of the arbitrariness of the term "disease," indicating that in reality, there are no diseases and only specific sick people, but it is not possible for him to make a precise distinction between the concrete and the abstract,[19] as is the strict application of the terms "disease" or "health."[20]

What is more, he claimed that it is impossible to make a sharp distinction between normality and abnormality, considering that very often there is no difference in quality, but only in quantity.[21] It seems that we tend to call pathology what does not fit into what is standard for our thought style.[22] The unification of any concept in one thought style occurs only due to a "coupling of phenomena resulting from historical and cultural circumstances."[23] This coupling is not treated as a metaphysical necessity but as a historically established convergence of multiple phenomena. However, we cannot manipulate freely as individuals, as it is constituted on a collective level and is socially reinforced in its further development. All forms of thinking, which we call a priori or intuition, are elements of the collective nature of knowledge and can be studied by methods of the sociology of thinking.[24]

As has already been said, thinking and cognition can only occur within a community that functions within a specific culture, and socio-cultural factors

---

18 Fleck, *Style myślowe i fakty*, p. 55.
19 Fleck, *Psychosocjologia poznania naukowego*, p. 54, fn. 1.
20 Fleck, *Psychosocjologia poznania naukowego*, p. 87.
21 Fleck, *Psychosocjologia poznania naukowego*, pp. 67, 83. As a practicing bacteriologist, he also claimed that it is impossible to draw an acute line between the physiological and the pathological in the biological sense.
22 Fleck, *Style myślowe i fakty*, p. 261.
23 Fleck, *Psychosocjologia poznania naukowego*, p. 42.
24 Fleck, *Style myślowe i fakty*, p. 180.

are a prerequisite for observation to occur at all (culture enables something). However, the same factors render observations that are alien to this culture impossible to be made at all (culture limits something). Fleck stresses that we always see through the eyes of the collective to which we belong.[25] Once it has been pulled out of the thought style in which it was created and perpetuated, any phenomena will become incomprehensible and meaningless. None of them can exist in isolation since none refers to something that would exist "in itself."[26] To paraphrase Fleck's saying, the truth about anorexia must be treated as a tripartite relationship between the subject (the judgments established about it), the "state of things"/object (described and constituted in a mental style), and the thought collective (which also includes the current state of knowledge and mental culture). In other words, what we define as truth is simply the only possible solution to a problem that is consistent with a thought style, its particular "compulsion."[27] In this perspective, anorexia as such does not exist. However, there is a phenomenon which, in the medical thought style (which seems to permeate the popular thought style for the reason of its dissemination and popularization) functions as pathology, and in feminist thought style, peripheral to the former, as an affliction of contemporary culture, embodied, not accidentally, mainly in women. It is worth mentioning that Fleck's conviction that almost all the theories developed from some ancient idea, which is not yet of a scientific nature and often came from a common, magical, or religious thought style. The continuity of historical understanding allows the researcher(s) of anorexia to indicate that the sources of our culture's susceptibility to the presence of anorexia can be justifiably found in the Christian-Cartesian dualism, in reducing the value of femininity to carnality or in the rationalistic understanding of subjectivity. There is no question of homogenous causality (Fleck calls this moment the "nucleus of development"), but instead of generating the conditions for a phenomenon to develop and be perpetuated in one form rather than another.[28] In this sense, human cognition is neither empirical nor speculative, but it results from the historical evolution of thought. It derives, as Fleck writes, from evolutione historica.[29] Fleck's psychosociology of scientific cognition, which is,

---

25 Fleck, *Style myślowe i fakty*, pp. 140–141, 169.
26 Fleck, *Psychosocjologia poznania naukowego*, p. 69; Fleck, *Style myślowe i fakty*, p. 61.
27 Fleck, *Psychosocjologia poznania naukowego*, p. 122; Fleck, *Style myślowe i fakty*, pp. 181, 188.
28 Fleck, *Psychosocjologia poznania naukowego*, pp. 55–56; Fleck, *Style myślowe i fakty*, p. 100.
29 Fleck, *Style myślowe i fakty*, p. 229.

by all means, a precursor, not only to the research of Thomas Kuhn but also to Michel Foucault (and his archaeology of knowledge), is filled with a beguiling, almost infectious optimism about the actions of humanity. Fleck believed that the adoption of his tripartite model of knowledge (subject-object-collective) would eliminate hypocrisy in science (because it would know its conditions and limitations), the natural sciences and the humanities would become closer to each other (because the former would not have privileged access to "reality"). Whereas, natural science itself "will become more human." It will also make it possible to narrow the existing gap between theory and scientific practice and increase the loyalty of scientists towards humanity.[30] The modern phenomenon of anorexia, as described and analysed by feminist thinkers, evokes both fear and optimism (it is difficult to say unequivocally in what proportion). They show that we often deal with a "dehumanized" manner of treating anorexic patients, which is particularly evident in psychiatric discourse. A prestigious medical style thought constantly influences common thinking, which results in the development of modern pharmacology and the commercialization of both medical and pharmacological research. At the same time, we observe attempts to "pull" the phenomenon of anorexia "out" of medical discourse, to demonstrate its multidimensionality, and to empower people suffering from anorexia as entirely rational and sane people.

Historical Fate of Anorexia

With great approximation, we can say that anorexia nervosa in its present known form was conceived in the 1870s and had two self-proclaimed fathers, Sir William Gull and E. C. Lasèque. The first was a well-known and respected British doctor in the second half of the nineteenth century, a lecturer in medicine, a member of the Royal Society, president of the Clinical Society, a consultant doctor of the famous Guy's Hospital in London, and after 1872 a doctor of Queen Victoria herself. The second was a recognized and valued doctor in France, an expert in women's ailments, who studied gastric disorders on a small group of women. They published their works almost simultaneously, Gull in 1874, the article "Anorexia nervosa (apepsia hysterica, anorexia hysterica)" (*Transactions of the Clinical Society of London*, 7/1874: 22–28), and Lasèque in 1873 "De l'anorexie hystérique" (*Archives Générales de Médecine* 21/1873: 385–403).

---

30 Fleck, *Style myślowe i fakty*, p. 189.

Both works are a good illustration of the historical transformation in Western culture, consisting of the increasing recognition of the authority of medicine and doctors instead of the authority coming from magical or religious beliefs. For a moment, then, let us go back in time. Anorexia is an old phenomenon and has been recorded in the literature for a very long time. Its triumph as a sign of sanctity, a miracle of life deprived of the bodily need to eat, was celebrated in the late Middle Ages (especially in the twelfth to fourteenth centuries).[31] Fascinating is the connection between sanctity, conscious refusal to eat, and women. The capacity to refuse to eat was an expression of admired asceticism, which allowed women to maintain their high social status and personal subjectivity. Many women who starved to death, such as Catherine of Siena, were considered saints by male religious authorities at the same time when femininity was increasingly identified with evil, deviating from the (male) norm, deceptive, and impure. Such identification would eventually materialize with the barbaric witch-hunt's spread, characteristic of the 15th and 16th centuries. It has several elements that will become characteristic of thinking about anorexia in later times. First of all, asceticism, contrary to the ancient tradition, consists of questioning corporeality, the abandonment of the corporeal urges of the body, and the "animal, unreasonable" physiology associated with it. Carnality, animality, and physiology are attributes long associated with femininity;[32] no wonder then that their resignation allowed women to regain their status as full-fledged subjects. Secondly, the ability to control hunger made it possible to control the body and permitted the spirit's victory (reason) over matter (body). It is reflected in the stories of contemporary anorexic women who seek to "dispose of" their bodies and gain complete control over them. Thirdly, abandoning the idea of "sacred anorexia" is a sign of a slow but consistent tracking of deviation (and not sanctity) in women's behaviour, which manifested itself both in their dealings with Satan and in their hysteria, mental instability, or excessive emotionality.

The motifs mentioned above lie in the original ideas, which will become the nucleus of the modern, medicalized understanding of anorexia. Early premedical descriptions of anorexia date back to the first half of the 17th century. Hieronim Fabricius describes the case of a thirteen-year-old girl who allegedly lived without drinking or eating for three years. He also describes her characteristic

---

31 Hepworth, *The Social Construction of Anorexia Nervosa*, p. 14.
32 David Gilmore, *Misogyny: The Male Malady*. Philadelphia: The University of Pennsylvania Press, 2009.

emotional state because she was sad and melancholic.³³ The first detailed, widely recognized report on anorexia by medical historians was compiled by Richard Morton and published as Phthisiologia or a treatise of consumptions around 1694. Morton presented women in a morbid state of mind who refused to eat. He described this phenomenon as nervous consumption. For Morton, consumption and phthisis (a Latin word derived from the Greek word, meaning wilting, becoming miserable) constituted certain states, which triggered the disease. He called it "nervous atrophy" and sought its causes in the violent passions of the mind. It aligned with the Georgian holistic understanding of man, with no distinction between psychological and somatic elements. It is worth emphasizing here that while Morton's description is medical, he uses physicians' discourse of his time; nevertheless, nervous atrophy is not a disease in today's meaning of the word. The cases described by Morton fall more under the fasting phenomenon than today's anorexia nervosa.³⁴ Shortly afterward, in Victorian times, women with similar symptoms were referred to as fasting girls.

In medical dictionaries, long before Gull's, the term anorexia nervosa was adopted, the term anorexia mirabilis (miraculous lack of appetite) was used.³⁵ Today this name is used almost exclusively for saints who refused to eat for reasons of faith. Lasèque used the term anorexie hysterique in his article. Gull considered that the terms anorexia hysterica, apepsia hysterica (apepsia: disorders of the functioning of the digestive system), which he still used as a primary term in a reading given a year earlier at a meeting of the Clinical Society, were less accurate than the term anorexia nervosa, which means a lack of appetite (anorexia, from Greek: "an"- lack, "orexis"- appetite). The adjective hysterica, present in the above expressions, is extremely important as it illustrates the established belief in Western culture that hysteria is a female affliction. The Greek word "hystera/hustera" stands for uterus. However, it quickly became a term that described the "odd," abnormal female behaviour that began to be associated with women's mental condition and emotionality. Hysteria, like hypochondria, were concepts commonly used in the 18th and 19th centuries to describe almost every kind of behaviour or perception that was classified as a hypersensitivity.³⁶ Thus, the word, which calls one of almost all women's organs (because they possess the

---

33 Malson, *The Thin Woman*, p. 51.
34 Malson, *The Thin Woman*, pp. 52–54.
35 Hepworth, *The Social Construction of Anorexia Nervosa*, p. 24.
36 The nineteenth century, especially its later years, was even referred to as the golden age of hysteria.

uterus), has become a metaphor to render the behaviour and mental proclivity typical for them. Women have become hysterical, at best emotional, and at worst mentally unstable. The adjective' nervosa' was used in the nineteenth century to describe many phenomena dealt with by the developing science called psychopathology, which confirmed the relationship between mind and body and the psyche's influence on the body (cf. Freud's discoveries). The replacement of the term "hysterica" with "nervosa" made anorexia a phenomenon dealt with by medicine, a specific type of nervous disease that could be legitimately examined and treated only with scientific methods. Women stopped being hysterical and became neurotic.[37] The anorexia presented in Gull and Lasèque's works is a mental disorder related to physiological predispositions, which occurs essentially in women. Due to the last aspect, both authors proposed that the described condition can be understood only after considering current knowledge about women and femininity. Thus, anorexia was considered in the context of 19th-century views on femininity, which generally equated femininity with irrationality, which resulted from the widespread attribution of hysteria to women, as mentioned above. Gull called anorexia a "mental perversion," which is caused by ego distortions. At the same time, Lasèque described the women suffering from it as "lunatics," seeking the cause of the disease in the emotional and unstable nature of women.[38] According to the understanding of the time, perversion was a kind of tenacity or stubbornness rather than deviation; as we understand it today, anorexia once and for all established its connection with a mental disorder in medical discourse. Both Gull and Lasèque believed that patients should be subjected to meticulous control (including moral control) by third persons (preferably qualified medical personnel). They should be fed regularly and by force, isolated from relatives or friends, which was characteristic of the psychiatric practice of their time.[39] The patients were treated as small children devoid of subjectivity, both mentally and morally weak, and unable to take responsibility for their actions.[40] Refusal to eat was thus linked to a particular group of people's characteristic emotional and mental syndromes, revealing the irrationality and abnormality attributed to them.

---

37 Malson, *The Thin Woman*, p. 48.
38 Hepworth, *The Social Construction of Anorexia Nervosa*, pp. 30, 33.
39 Hepworth, *The Social Construction of Anorexia Nervosa*, p. 36.
40 Malson, *The Thin Woman*, p. 65.

Anorexia Nervosa in Medical Thought Style: Individual Deviation

As in the case of other diseases, specific criteria have been adopted to diagnose anorexia nervosa. Two sources are the most known and commonly used: Diagnostic and Statistical Manual for Mental Disorders (DSM IV) and the International Classification of Diseases (ICD 10):

According to DSM IV, we are dealing with anorexia nervosa when the following occurs:

1. Refusal to maintain body weight at or above the minimally normal weight for age and height (<85 % of normal body weight);
2. Intense fear of gaining weight or becoming fat, even though underweight;
3. Disturbance in the way in which one's body weight or shape is experienced, undue influence of body weight or shape on self-evaluation, or denial of the seriousness of the current low body weight;
4. In postmenarcheal females, amenorrhea, i.e., the absence of at least three consecutive menstrual cycles.

As in the case of other diseases, specific criteria have been adopted for the diagnosis of anorexia nervosa. Two sources are the most known and commonly used: Diagnostic and Statistical Manual for Mental Disorders (DSM IV) and the International Classification of Diseases (ICD 10):

According to the International Classification of Diseases (ICD 10) published by the World Health Organization, anorexia nervosa is: "A disorder characterized by deliberate weight loss, induced and sustained by the patient. It occurs most commonly in adolescent girls and young women. However, adolescent boys and young men may also be affected, as children approaching puberty and older women up to menopause. The disorder is associated with specific psychopathology whereby a dread of fatness and flabbiness of body contour persists as an intrusive, overvalued idea, and the patients impose a low weight threshold on themselves. There is usually undernutrition of varying severity with secondary endocrine and metabolic changes and disturbances of bodily function. The symptoms include restricted dietary choice, excessive exercise, induced vomiting and purgation, and use of appetite suppressants and diuretics" (F 50.0). As we can see, the medical discourse qualifies anorexia nervosa as a mental and behavioral disorder or a mental and behavioral syndrome or as a psychosomatic disorder with a complex etiology and individualized course. DSM concentrates rather on psychological aspects, whereas the ICD focuses on behavioural dimensions.

The causes of anorexia are sought by medicine in genetic predisposition,[41] hormonal disorders, affectivity disorders expressed in fear of puberty,[42] a cognitive dysfunction consisting mainly in a disturbed perception of one's own body, associated with widely disturbed psychological functioning, like for example lack of fully developed sense of the self,[43] Ultimately, it is a type of mental illness that has no reasonable explanation. In the psychological literature on the subject, quite often it is stated that anorexia is a pathological condition connected with bodily discomfort, emotional tension and visible long-term duration of stress.

The status of anorexia as a negative (to be eliminated) pathological phenomenon was confirmed here, and a group of experts (physicians) was being established who would be able to remedy it using scientific methods and identifying its causes. It would not be possible to analyse anorexia by physicians and medicine without establishing the institution of the hospital, subjects called patients, or the legitimacy of clinical research, which are the key elements that allow the medical thought style collective to function.[44] At this point, we cannot fail to mention that the paternalistic dimension of practices of psychiatric hospitals, especially in the 19th century and in the first half of the 20th century, led to the use of lobotomies, leukotomies, insulin shocks, or electroshocks in the treatment of mental disorders (including anorexia, schizophrenia, or depression).[45] Most psychiatric patients of that time were women.

---

41 Wade Berrettini, "The Genetics of Eating Disorders," *Psychiatry (Edgmont)* 1(3)/2004, pp. 18–25.
42 Monika Olejniczak, "Anorexia nervosa and psychosexual functional of girl," *Psychiatria* 9(3)/2012, pp. 77–82.
43 National Institute for Clinical Excellence, *Eating Disorders. Core Interventions in the Treatment and management of Anorexia Nervosa, Bulimia Nervosa, and Related Eating Disorders*. London: Royal College of Psychiatrics, 2004; Janet Treasure, Ulrike Schmidt, Eric van Furth, *The Essential Handbook of Eating Disorders*. West Sussex: Wiley, 2005.
44 I refer the reader to excellent and exciting studies on the establishment of the clinic and the constitution of patients. Michel Foucault, *The birth of the clinic* (London: Routledge, 2003), Jürgen Thorwald, *The Patients* (Harcourt Brace Jovanovich, 1972). I also refer to Charlotte Perkins Gilman's moving book *The Yellow Paper*, in which she describes the typical fate of a nineteenth-century woman considered mentally ill.
45 The turn that took place in the theories on the origin of mental illnesses in the 1960s led to the emergence of the so-called anti-psychiatry; it is worth mentioning here the works of such researchers as Thomas Shash, Erving Hoffman, or Richard. D. Laing.

Anorexia Nervosa in Feminist Thought Style: A Non-Random Affliction of Modern Culture

It has been repeatedly mentioned that feminist studies pay special attention to the history of anorexia and its cultural entanglement. They demonstrate, following Fleck, that it cannot be solely treated as a medical entity. It should rather be seen as a result of the medical coupling of the discourse on hysteria (well established in the ordinary minds of the 19th century) with the results of the research on gastric disorders developed on a nervous background.[46] When the history of anorexia is presented, its discursive dimension and its position in social and historical conditions become apparent. Thus, we realize that it cannot be understood only from the perspective of a single research discipline (e.g., medicine). The multidisciplinary approach is needed, in which it is assumed from the start that anorexia is a highly complex phenomenon, which, moreover, is not questioned nowadays in medical style either (cf. the WHO qualification above). Once again, it should be emphasized that anorexia (as a phenomenon and its associated interpretation) does not exist independently of the discourses in which it is constructed and subjected to consideration; scientific, clinical, informal, and popular media discourses. Susan Bordo emphasizes that in feminist reflection on anorexia, we undertake a difficult task because we have to convince academics and doctors (or other so-called professionals) that the creations of cultural imagination and the effects of their actions must be taken seriously. They should be considered as potential causes of the described phenomena.[47]

It seems that Sigmund Freud's psychoanalysis and its later reinterpretations have made a way to less narrow approaches to anorexia. They overcome the physiological thinking of medicine and no longer look for the causes of the disease solely in biological or physiological factors. Instead, the psychological sources of the disease were examined, starting to treat anorexia nervosa more as a sociocultural phenomenon and not the individualistic one. In this way, systemic family theories were developed, postulating research on the psychosexual development of anorexic patients, their families, and their relationships. In this perspective, anorexia was described as an illusory solution to problems related to personality and interpersonal relations. Particularly innovative was the therapeutic approach of Hilde Bruch, known in feminist literature. She called anorexia a kind of defensive response to the sense of a lack of a person's personality and its

---

46 Malson, *The Thin Woman*, p. 67.
47 Bordo, *Unbearable Weight*, p. xxi.

foundations: the sense of being powerless and incompetent.[48] She was one of the first to describe and name the anorexia-related disorder of perception of one's body size and called it "disturbance in size awareness." It became a commonly accepted hallmark of anorexia, both in medical literature and in popular belief. Seeking the sources of the disease in the family resulted in a thorough analysis of the mother-daughter relationship (which was indirectly the result of psycho-analytical tradition), the mother's role, and her influence on the development of nutritional patterns in the family. Mother-daughter relationships were linked to subjectivity, individual identity, and the self-confidence of both mothers and daughters. The methods themselves also changed, as the patients' narrative and how they reported and conceptualized their situation became important.[49] Mothers are the crucial actors in the literature on psychoanalytic approaches to anorexia. They are treated both as the cause of the development of the disease and as a source of explanation (the so-called "blame the mothers' issue"). The apparent absence of the father and other social factors (where social is understood more broadly than simply the family) became the subject of feminist criticism, especially in the 1990s.[50] It seems that in the described psychological modification, the categorization of anorexia as a pathology remained intact; it ceased to be strictly an individual pathology and became a family pathology.[51] It still lacked the introduction of a broader socio-cultural context and a reflection that most patients were women.

The methodological postulate of feminist research on anorexia, which prohibited the treatment of women diagnosed as anorexics as completely different (sick, pathological, deviant) from so-called ordinary women, proved to be highly prolific. Feminists raised the question: Perhaps anorexic women are not that different from other women? Moreover, an attempt was made to explain a specific cultural pattern in the contemporary approach to eating, which can be presented in a sequence of terms: normal eating-diet-anorexia. Thus, an essential link between a slim female figure's cultural idealization, paranoia regarding weight loss, and the increasing number of eating disorders was noted.[52] Being on a diet or expressing the need to be on a diet and dissatisfaction with the appearance

---

48 Hilde Bruch, "Anorexia Nervosa: Therapy and Theory," *American Journal of Psychiatry* 139/1982, pp. 15–32.
49 Chernin, *The Hungry Self*; Orbach, *Fat is a Feminist Issue*; MacLeod, *The Art of Starvation*.
50 Hepworth, *The Social Construction of Anorexia Nervosa*, p. 39.
51 Malson, *The Thin Woman*, p. 88.
52 Malson, *The Thin Woman*, pp. 5, 90.

of one's own body, given their prevalence among girls and women, seem much more "normal" than not being on a diet. Thus, anorexia cannot be treated as an affliction that can be separated from women and girls' whole experience connected with eating. This experience is much more than ever before in our culture, associated with the media coverage of feminine beauty (the ideal size is becoming smaller and smaller) and a widespread belief in fat and obese people's ugliness and incapability.[53] It is no accident that the increase in anorexia's cases coincides with the growing obsession with appearance in our culture, which is expressed in the desire to have a slim figure (like the male and female models known from the media), as well as in the constant fear of possible weight gain. It is confirmed that weight-related concerns were not recorded almost until the 1960s in patients diagnosed as anorexic. Furthermore, anorexia is predominantly a matter of Western culture and appears in those communities under its strong influence.[54] With the support of modern biotechnology, cosmetology, and aesthetic surgery, today's bodies increasingly feed on fantasies of transformation and correction and permanent modification. Their materiality is almost denied.[55] The fact that it is women who dominantly suffer from anorexia can legitimately be associated with a culturally established conviction that female subjectivity (much more so than male subjectivity) is associated with physical appearance. Thus, although the media message also affects men, physical perfection standards (including those of slimness) concern women more than men. To them, Foucault's theses about the body as a place of social control apply above all.[56] The research conducted on women suffering from anorexia, in particular the interviews conducted, vividly showed that these women do not feel good in their bodies and with their bodies. It is known that the less one accepts one's own body, the more one is inclined to have low self-esteem, including exaggerating the body's size.[57]

---

53 Salas Ramos, Forhan M, Caulfield T, Sharma AM and Raine KD, "Addressing Internalized Weight Bias and Changing Damaged Social Identities for People Living With Obesity." *Front. Psychol.* 2019. https://www.frontiersin.org/articles/10.3389/fpsyg.2019.01409/full (20 Dec. 2021).
54 Malson, *The Thin Woman. Feminism, Post-structuralism and the Social Psychology of Anorexia Nervosa*, p. 93.
55 Bordo, *Unbearable Weight*, pp. 245–275.
56 Sandra Bartky Lee, *Foucault, Femininity, and the Modernization of Patriarchal Power* (London: Routledge, 1990).
57 Marika Tiggemann, "Body dissatisfaction and adolescent self-esteem: Prospective findings," *Body Image* (2)/2005, pp. 129–135; L.Parnot, Rousseau A., Benazet M., Faure B., Lenhoret E., Sanchez A., Chabrol H. "Comparaison du vécu corporel en fonction

Gender, therefore, seems to be the primary factor in the onset of the phenomenon of anorexia nervosa and its so many reproductions, and hence cannot be treated as a mere accompanying factor.[58] An anorexic in this perspective is someone who has too carefully assimilated cultural standards of appearance and applies them too precisely, and not someone who misperceives his or her own body.[59] The sick women are not irrational beings who cannot correctly relate to "external" reality, living in the mythological world of their distorted beliefs, because they can read socially established demands for slenderness or "being on a diet."[60] Paraphrasing the aforementioned psychological term of anorexia, one can say that the conditions in which women have to build their subjectivity are pretty unfavorable. However, anorexics can almost perfectly adapt to them by literally meeting cultural expectations.[61] Starving (similarly to overeating) has become our times' ideology and was not produced by the individual invention of women suffering from anorexia. Due to their biologically and culturally driven sensitivity about their bodies, women have become victims of this ideology.

The feminist study of the anorexigenic nature of culture is not limited to analyzing phenomena occurring at the level of popular culture, of which the most widely disseminated mass media are a good example. Researchers also looked very closely at the traditional understanding of subjectivity and the relations among femininity, body, and power.[62] To say that in modern culture, we are obsessed with a slim, beautiful body is not enough. Much more important seems to be the answer to the question, where did this obsession come from, and what constantly fuels or deepens it and causes it to affect women severely? To give it, Bordo uses three categories, three variables: the variable of dualism, the variable

---

du genre chez l'adolescent et le jeune adulte," *Journal de Thérapie Comportementale et Cognitive* 16 (2)/2006, pp. 45–48; Anne Morris and Debra Katzman. "The impact of the media on eating disorders in children and adolescents," *Paediatrics & child health* 8(5)/2010, pp. 287–289; Trisha Pruis, Jeri S Janowsky, "Assessment of Body Image in Younger and Older Women," *The Journal of General Psychology* 137 (3)/2010, pp. 225–238; D. Mellor, M. Fuller-Tyszkiewicz, M. P. McCabe, "Body Image and Self-Esteem Across Age and Gender: A Short-Term Longitudinal Study," *Sex Roles* 63/2010, pp. 672–681.

58 Bordo, *Unbearable Weight*, p. 54.
59 Bordo, *Unbearable Weight*, p. 57.
60 Bordo, *Unbearable Weight*, p. 59; R. N Ata, A. B Ludden and M.Lally, "The effects of gender and family, friend, and media influences on eating behaviors and body image during adolescence," *Journal of Youth and Adolescence* 36(8)/2007, pp. 1024–1037.
61 Helen Malson and Maree Burns, eds. *Critical Feminist Approaches to Eating Dis/Orders*.
62 White, *The Spirituality of Anorexia*.

of control, and the variable of cultural gender and power, which intersect in the phenomenon of anorexia.[63] The dualism of soul/reason/mind (depending on the variant) and body is a centuries-old heritage of our culture, which manifests itself in many dimensions. Let us simplify them here for the sake of the article. The basis of subjectivity or constitution of identity is the reason (mind); the body is something alien, a non-self. The body is an obstacle, a prison, a trap for a reason (soul); one must free oneself from it. It is an enemy that deforms thinking and threatens reason, hindering self-control, an idea which resounds in the texts of Plato, Augustine, Descartes, and many others. Therefore, it is necessary to develop procedures and a strategy through which reason will be able to conquer the body and subject it to meticulous control. All the factors mentioned above are repeatedly visible in the accounts of anorexic patients. Most of them claimed that losing weight gives them a sense of taking control of the body, themselves, achieving a state of cleanliness or specific perfection, and having a strong will of being better than others. All of this makes it possible to build one's own identity, to form a subjectivity. Many patients with anorexia wanted to reach the point where they would no longer have a body but achieve a state of imaginary total disembodiment (the quintessence of subjectivity). Cultural patterns of femininity play an essential role here. Most patients manifest a characteristic fear of adolescence, of having an entire female body (a symbol of feminine sexuality), becoming a woman: they dream of a child's body, which would not grow breasts and whose hips would not become rounded. Many are happy that they stop menstruating, a bodily function that is considered quite commonly an indicator of femininity (in medical style as one of the diagnostic criteria). This fear is most often explained by the fear of entering into sexual life (often caused by some trauma from childhood), of entering into socially imposed female roles of mothers, wives, guardians. However, it is also interpreted as a peculiar, exaggerated protest against archetypal femininity images as an uncontrolled emotionality that cannot be tamed. The anorexic seems to be an example for herself of mastering control to the perfect degree, which gives her a sense of power over herself, the power she needs to build her subjectivity, such power that the culture in which she has to live persistently denies her.[64]

---

63 Bordo, *Unbearable Weight*, p. 142.
64 Bordo, *Unbearable Weight*, pp. 154–164.

Instead of a Happy Ending

To sum up, the medical collective wishes to see anorexia as individual psychopathology, whereas in the feminist collective, the disease is created due to complicated socio-cultural processes and concentrated all that has gone wrong in this culture. For if all the power and energy that anorexics put into achieving the ideal they anticipate from cultural reality is an example of a kind of feminist protest, then, as Bordo points out, the obsession with slimness is hopelessly unproductive and ineffective. It seems that anorexia nervosa is still not fully standardized in medical discourse and still misses many actors from its view.[65] Moreover, we can observe the growing role of body appearance for young adolescents (both girls and boys) due to the impact of the contemporary mass media and social media.[66] Hence one can speculate that the phenomenon of eating disorders (including anorexia nervosa) will remain vital and interesting in western culture. Therefore we should redouble our efforts to join conceptual forces from the medical thought style and feminist thought style on anorexia I referred to in this article. Generally speaking, the phenomenon of disease can be explored at the intersections of physical, organismic aspects (the domain of medicine and biology), the patient's human sociocultural endeavour (the holistic approach in medicine, critical socio-cultural studies), and the existential dimension of the affected person (philosophy, theology, narrative studies). In other words, anorexia nervosa as a multi-layered phenomenon cannot be captured by a single scientific domain or theory. We must integrate efforts from socio-culturally and naturalistically oriented theories to help the suffering patients and change our culture's worthless obsession with the attractiveness of the individual body.

Bibliography

Ata, R. N., Ludden, A. B., & Lally, M. "The effects of gender and family, friend, and media influences on eating behaviors and body image during adolescence." *Journal of Youth and Adolescence* 36(8)/2007, pp. 1024–1037.

---

[65] Maroń, "Standardization of Medicine – case of anorexia nervosa," p. 228.
[66] Morris and Katzman, "The impact of the media on eating disorders in children and adolescents," Annalise Mabe, Jean K Forney and Pamela Keel, "Do you 'like' my photo? Facebook use maintains eating disorder risk," *International Journal of Eating Disorders* 47 (5)/2014, pp. 516–523; Christopher Ferguson, Monika Muñoz and Adolfo Garza, "Concurrent and Prospective Analyses of Peer, Television and Social Media Influences on Body Dissatisfaction, Eating Disorder Symptoms and Life Satisfaction in Adolescent Girls," *Journal of Youth and Adolescence* 43/2014, pp. 1–14.

Bartky Lee, Sandra. *Foucault, Femininity, and the Modernization of Patriarchal Power*. London: Routledge, 1990.

Berrettini, Wade. "The Genetics of Eating Disorders" *Psychiatry (Edgmont)* 1(3)/2004, pp. 18–25.

Birmingham C. Laird, Su Jenny, Hlynsky Julia A., Goldner Elliot M., Gao Min. "The mortality rate from anorexia nervosa." *Eating disorders* 38 (2)/2005, pp. 143–146.

Bordo, Susan. *Unbearable Weight: Feminism, Western Culture and the Body*. University of California Press: London, 2003.

Bruch, Hilde. "Anorexia Nervosa: Therapy and Theory." *American Journal of Psychiatry* 139/1982, pp. 1531–1538.

Chernin, Kim. *The Hungry Self: Daughters and Mothers, Eating and Identity*. Virago Press: London, 1986.

Chernin, Kim. *The Obsession: Reflections on the Tyranny of Slenderness*. New York: Harper and Row Publishers, 1981.

Ferguson, Christopher, Muñoz Monika, Garza Adolfo et al. "Concurrent and Prospective Analyses of Peer, Television and Social Media Influences on Body Dissatisfaction, Eating Disorder Symptoms and Life Satisfaction in Adolescent Girls." *Journal of Youth and Adolescence* 43/2014, pp. 1–14.

Fleck, Ludwik. *Genesis and Development of a Scientific Fact*. Chicago: University of Chicago Press, 1979.

Fleck, Ludwik. *Psychosocjologia poznania naukowego. Powstanie i rozwój faktu naukowego oraz inne pisma z filozofii poznania*. Lublin: UMCS, 2006.

Fleck, Ludwik. *Style myślowe i fakty. Artykuły i świadectwa*. Warszawa: IFiS PAN, 2007.

Foucault, Michel. *The birth of the clinic*. London: Routledge, 2003.

Franko, Debra L, Keshaviah Aparna, Eddy, Kamryn T., Krishna, Meera, Davis, Martha C., Keel, Pamela K., Herzog David B. "A Longitudinal Investigation of Mortality in Anorexia Nervosa and Bulimia Nervosa." *The American Journal of Psychiatry* 2013. https://doi.org/10.1176/appi.ajp.2013.12070868 (20 Dec. 2021)

Gilmore, David. *Misogyny: The Male Malady*. Philadelphia: The University of Pennsylvania Press, 2009.

Hall, Lindsey, Ostroff, Monika. *Anorexia nervosa. A Guide to Recovery*. Carlsbad: Gurze books, 1999.

Hepworth, Julie. *The Social Construction of Anorexia Nervosa*. London: Sage Publications, 1999.

Mabe Annalise, Forney Jean K., Keel Pamela K. "Do you 'like' my photo? Facebook use maintains eating disorder risk." *International Journal of Eating Disorders* 47 (5)/2014, pp. 516–523.

MacLeod, Sheila. *The Art of Starvation*. London: Virago, 1981.

MacSween, Morag. *A Feminist and Sociological Perspective on Anorexia Nervosa*. London: Routledge, 1993.

Malson, Helen and Maree Burns, eds. *Critical Feminist Approaches to Eating Dis/Orders*. London: Routledge, 2009.

Malson, Helen. *The Thin Woman. Feminism, Post-structuralism and the Social Psychology of Anorexia Nervosa*. London: Routledge, 1998.

Maroń, Piotr. "Standarization of Medicine – case of anorexia nervosa." In: *Graszewicz.com. Media. Komunikacja. Kultura* [*Graszewicz.com. Media. Communication. Culture*], ed. Dominik Lewiński, Karina Stasiuk-Krajewska, Roman Wróblewski. Wrocław: Libron, 2017, pp. 211–229.

Mellor, D., Fuller-Tyszkiewicz, M., McCabe, M.P. et al. "Body Image and Self-Esteem Across Age and Gender: A Short-Term Longitudinal Study." *Sex Roles* 63/2010, pp. 672–681.

Morris, Anne, Katzman, Debra. "The impact of the media on eating disorders in children and adolescents." *Paediatrics & child health* 8(5)/2010, pp. 287–289.

National Institute for Clinical Excellence. *Eating Disorders. Core Interventions in the Treatment and management of Anorexia Nervosa, Bulimia Nervosa, and Related Eating Disorders*. London: Royal College of Psychiatrics, 2004.

Olejniczak, Monika. "Anorexia nervosa and psychosexual functional of girl." *Psychiatria* 9(3)/2012, pp. 77–82.

Orbach, Susie. *Fat is a Feminist Issue*. London: Hamlyn, 1978.

Orbach, Susie. *Hunger Strike: The Anorectic's Struggle as a Metaphor for Our Age*. London: Faber and Faber, 1986.

Overbeke, Grace. "Pro-Anorexia Websites: Content, Impact, and Explanations of Popularity." *The Wesleyan Journal of Psychology* 3/2008, pp. 49–62.

Parnot, L. Rousseau A., Benazet M., Faure B., Lenhoret E., Sanchez A., Chabrol H. "Comparaison du vécu corporel en fonction du genre chez l'adolescent et le jeune adulte." *Journal de Thérapie Comportementale et Cognitive* 16 (2)/2006, pp. 45–48.

Pruis Trisha, Janowsky Jeri S. "Assessment of Body Image in Younger and Older Women." *The Journal of General Psychology* 137 (3)/2010, pp. 225–238.

Ramos Salas X, Forhan M, Caulfield T, Sharma AM and Raine KD. "Addressing Internalized Weight Bias and Changing Damaged Social Identities for People

Living With Obesity." Front. Psychol. 2019. https://www.frontiersin.org/articles/10.3389/fpsyg.2019.01409/full (20 Dec. 2021).

Swain, Pamela I. ed. *Anorexia nervosa and bulimia nervosa: New research*. New York: Nova Science Publishers, 2006.

Thompson, Becky. *A Hunger So Wide And So Deep: A Multiracial View of Women's Eating Problems*. Minneapolis: University of Minnesota Press, 1994.

Thorwald, Jürgen. *The Patients*. Harcourt Brace Jovanovich, 1972.

Tierney, Stephanie. "The dangers and draw of online communication: Pro-anorexia websites and their implications for users, practitioners, and researchers." *Eating Disorders* 14 (3)/2006, pp. 181–190.

Tiggemann Marika. "Body dissatisfaction and adolescent self-esteem: Prospective findings." *Body Image* 2/2005, pp. 129–135.

Treasure Janet, Schmidt Ulrike, van Furth Eric. *The Essential Handbook of Eating Disorders*. West Sussex: Wiley, 2005.

White, Emma. *The Spirituality of Anorexia: A Goddess Feminist Thealogy*. London: Routledge, 2019.

Ewa Bińczyk
Translated by Szymon Włoch

The Disinformation Rhetoric of the Twenty-First Century[1]

Abstract: The text presents the most prevalent rhetorical techniques used in the so-called global warming denial movement, especially in the USA. It focuses on such methods of disinformation used by merchants of doubt as: "there is no proof," "paralysis through analysis," greenwashing/aggressive mimicry, and the "junk science" accusations. The last part of the text asks about the current possibilities to rebuild the authority of science in the epoch of professional denialism on demand.

Keywords: climate change, disinformation, denialism, rhetoric, greenwashing, authority of science

We are jeopardizing our future.[2]

Down the Garden Path

Since 2008, the United States of America rank second (behind China) in national greenhouse gas emission. The next spots on this odious list belong to Indonesia and Brazil.[3] In 2006, 35 % of Americans believed that global warming is not caused by human activity but is a natural trend. In 2007, about 40 % of America's population was convinced that scientists still cannot agree on whether the global warming we are experiencing is caused by human activity.[4] In 2009,

---

1 The text is an English and modified version of the chapter number 7 of the book *Epoka człowieka. Retoryka i marazm antropocenu* (Warszawa: Wydawnictwo Naukowe PWN, 2018).
2 William J. Ripple, Christopher Wolf, Thomas M. Newsome et al., "World Scientists' Warning to Humanity. A Second Notice." *BioScience* 67(12)/2017, p. 1026.
3 Rasmussen Reports, "Energy Update," 2012. (10 October 2013, currently website is not available) http://www.rasmussenreports.com/public_content/politics/current_events/environment_energy/energy_update, Nicholas Stern, *The Global Deal. Climate Change and the Creation of a New Era of Progress and Prosperity* (New York: Public Affairs, 2009), p. 13.
4 Naomi Oreskes, Erik M. Conway, *Merchants of Doubt. How a Handful of Scientists Obscured the Truth on Issues from Tobacco Smoke to Global Warming* (New York: Bloomsbury Press, 2010), p. 241.

the percentage of people in that country who believed that global warming is not a product of human activity had risen to 44 %.[5] Not much had changed by 2011: 35–40 % of Americans still refuted that we are experiencing climate change or believed that the change is caused by humans.[6] How is this possible?

In 2007, 71 % of Americans believed that continued combustion of fossil fuels will lead to climate change. Just one year later, that same belief was reported by only 59 % of the people polled.[7] By 2011, that percentage had dropped further still, to 44 %. As the director of the Pew Research Center commented, that was one of the most dramatic public opinion shifts ever noted in such a short span of time.[8] So, what was reason?

In Anthony Giddens' 2009 book *The Politics of Climate Change* we read that skeptics have a shocking and disproportional degree of influence on public opinion regarding climate change – especially if we take in to account how few of the skeptic opinions of this kind actually come from climatologists.[9] Thanks to recent research in the field of science and technology studies, we can today answer the question of how this happened.[10] Because lurking behind the

---

5   Rasmussen Reports, "44 % Say Global Warming Due To Planetary Trends, Not People," research conducted by Rasmussen LLC on the 19th of January 2009 (10 October 2013, currently website is not available), http://www.rasmussenreports.com.
6   The Pew Research Center, "Modest Rise in Number Saying There is 'Solid Evidence' of Global Warming," 2011. (2 November 2015) http://www.people-press.org/files/legacy-pdf/12-1-11%20Global%20warming%20release.pdf.
7   The Pew Research Center, "Fewer Americans See Solid Evidence on Global Warming," 2009. (2 November 2015) http://pewresearch.org./pubs/1386/cap-and-trade-global-warming-opinion.
8   Cf. Naomi Klein, This Changes Everything. *Capitalism vs. The Climate* (New York: Simon & Schuster, 2014), p. 51.
9   Anthony Giddens, *The Politics of Climate Change* (Malden, MA: Polity Press, 2009), p. 101.
10 In his review essay, Hamilton rips Giddens' book to shreds, Clive Hamilton, "Theories of Climate Change," *Australian Journal of Political Science* 47(4)/2012, p. 726 ff. He determines it to be Eurocentric (as the author overstates the EU's role in global environmental politics) and repeatedly cites the British sociologist's lack of qualifications in climatology and Earth System science. Hamilton also states that Giddens does not understand the seriousness of the threat of climate destabilization. Giddens normalizes the dangers of climate change, thinking of them as one of many topics in his thought on political theory. Meanwhile, Hamilton believes climate change puts a question mark on further economic growth, world peace, and even the possibility of continued technological development. Hamilton believes the narrative in *The Politics of Climate Change* demonstrates a grave unfamiliarity with the debates on the causes of the climate policy

The Manufacture of Skepticism on Demand: The Global Warming Denial Movement

Science and Technology Studies literature describes many cases of spectacular disinformation campaigns – campaigns intended to produce doubt on demand. The most prominent examples were campaigns undermining the dangers of smoking: the correlation between smoking tobacco and lung cancer, nicotine being an addictive substance, and the dangers of second-hand smoke.[11] In other instances, such disinformation campaigns were used to breed controversy related to problems like acid rain, the danger of asbestos, holes in the ozone layer, and the negative effects of CFCs on the ozone layer. There has been systematic doubt-mongering surrounding the introduction of specific pro-environmental regulations in which not only the credibility of the IPCC was put in question but also that of the Environmental Protection Agency (EPA), existing in the United States since 1970. Also compromised was academic criticism towards projects like "Star Wars" and the anti-nuclear defense shield in US airspace (Strategic Defense Initiative, SDI), in which the credibility of the nuclear winter hypothesis was undermined.[12] Finally, there was doubt manufactured surrounding the

---

fiasco and the role of denialism (according to him, Giddens falls prey to the denialists on the matter that there is even a spectrum of serious and diverse stances in the debate on the causes of global warming). The British sociologist, Hamilton argues, does harm by marginalizing the political role of pro-environment movements and banalizing the premises behind low-emissions policies. Finally, Hamilton accuses the author of *The Politics of Climate Change* of being immodest for using the term "Giddens paradox" for the practice of discounting the future in the context of the climate change problem. He argues that this mechanism is so amply described in literature (which, coincidentally, the British sociologist does not cite) that it does not need a new name.

11 In 1992, about 25 % of Americans still did not believe that smoking is a health hazard. In 2007, 25 % of Americans were not convinced that we possess scientific evidence supporting the hypothesis that smoking can kill, Oreskes, Conway, *Merchants of Doubt*, pp. 33, 24.
12 One of the first people to issue a public warning of the possibility of nuclear winter as a consequence to the star wars program was the American astronomer Carl Sagan. In 1986, 6500 scientists signed a letter of protest against the American SDI program.

validity of the decision to ban the pesticide DDT, cited in which were economic and civilizational damages that would result from the pro-environmental law.[13]

One of the better-substantiated works on the stoking of skepticism on climate change is the empirical study *The Merchants of Doubt: How a Handful of Scientists Obscured the Truth on Issues from Tobacco Smoke to Global Warming*.[14] To a great extent, the book relies on data that saw the light of day with the release of court documents related to the class action lawsuit against the tobacco industry in the United States in the 1990s. Made available by the University of California, San Francisco as the *Legacy Tobacco Document Library*,[15] these documents made it

---

13 In 1955–1969, DDT was widely used in developing countries as part of the Global Malaria Eradication Campaign to combat insect-borne diseases like malaria. DDT was banned from use in the USA in 1972, as a result of discussions stimulated by the publication of one of the first works to openly address the problem of environmental contamination. The work in question is Rachel Carson's *Silent Spring* (Boston, MA: Houghton Mifflin, 1962), in which the author points out the possibility of long-term risks associated with DDT. Carson argued that the substance endangered other insects and could undergo bioaccumulation in the food chain, which could have a negative effect on the populations of certain bird species (like hawks and eagles). Around the year 2007, Carson was the target of extensive online attacks from critics of the regulation to ban DDT. It was argued that the decision to ban DDT caused the death of millions of people suffering from malaria. Carson was compared to Adolf Hitler. Naomi Oreskes and Erik M. Conway believe this was a deliberate disinformation campaign aiming to undermine the validity of introducing environmental regulations in general. These authors also point out that the Global Malaria Eradication Campaign failed to bring the desired results, anyway – as it later occurred, within seven to ten years after DDT spraying, most insects develop a resistance to the pesticide. In fact, the possibility of insects developing resistance to DDT was already postulated in 1947. Today, we have at our disposal extensive research data attesting to the dangers of DDT, including to humans. Global use of DDT was banned only in 2007, when Mexico stopped producing the pesticide. Involved in the campaign to generate doubt regarding the validity of banning DDT were institutions and individuals like: the Competitive Enterprise Institute, the Heartland Institute (sponsored by the Philip Morris Corporation), the Cato Institute, Steven Milloy and many media outlets (including The Wall Street Journal), Oreskes, Conway, *Merchants of Doubt*, pp. 216–239. All of these parties were also directly involved in the movement to negate global warming.

14 Oreskes, Conway, *Merchants of Doubt*. The book was the product of five years of historical research by the two authors. They analyzed "hundreds of thousands of document pages." Like the other few scientists who endeavored to expose the disinformation campaigns discussed in their book, Oreskes and Conway faced public attacks, Oreskes, Conway, *Merchants of Doubt*, pp. 242, 274, 264.

15 Available at: http://legacy.library.ucsf.edu (3 September 2021).

possible to connect the dots between the tobacco industry and the fuel industry, as well as certain scientific institutions and think tanks, and to reconstruct the working mechanisms of campaigns to undermine the objectivity of facts on demand.[16] We know today that representatives of the tobacco industry were aware of the carcinogenic effect of cigarette smoking as far back as the 1950s. In the American government's court case against the tobacco companies (USA vs Philip Morris Inc. et al.) the ruling was made on the basis of the fact that these companies concealed their knowledge of the situation.[17]

The documents in question are evidence of the fact that disinformation campaigns surrounding the dangers of smoking cigarettes and climate change were carried out according to a similar strategy. Inconvenient scientific findings were disputed by hired experts, lawyers and PR specialists. Among the documents contained in the *Legacy Tobacco Document Library* we find handbooks with instructions on how to manufacture doubt and controversy around a given thesis, as well as letters from experts ready to serve as commentators for hire and to provide desired testimony.[18] These documents also contain information on the effectiveness of various means to undermine inconvenient opinions from government agencies such as the Environmental Protection Agency. These pertain mainly to ways of undermining the scientific credibility of experts.

The documents in question are evidence of the fact that disinformation campaigns surrounding the dangers of smoking cigarettes and climate change were carried out according to a similar strategy. Inconvenient scientific findings were disputed by hired experts, lawyers and PR specialists. Among the documents contained in the *Legacy Tobacco Document Library* we find handbooks with instructions on how to manufacture doubt and controversy around a given thesis, as well as letters from experts ready to serve as commentators for hire and

---

16 On the phenomenon of global warming negation, see e.g.: Haydn W. Washington, John Cook, *Climate Change Denial. Heads in the Sand* (London: Earthscan Press, 2011); Hugh P. McDonald, "Denying Climate Change – Still!," *Capitalism, Nature, Socialism* 22(2)/ 2011, pp. 125–128; Steven A. Kolmes, "Climate Change. A Disinformation Campaign," *Environment. Science and Policy for Sustainable Development* 53(4)/2011, pp. 33–37, and the discussion: Steven Yearley, David Mercer, Andrew Pitman, Naomi Oreskes, Erik M. Conway, "Perspectives on Global Warming. Book Symposium," *Metascience* 21(1)/2012, pp. 1–29. Compare also Union of Concerned Scientists (Union of Concerned Scientists, 2007), and the Skeptical Science website: http://www.skepticalscience.com/ (3 September 2021).

17 Cf. Oreskes, Conway, *Merchants of Doubt*, pp. 31–33.

18 Oreskes, Conway, *Merchants of Doubt*, p. 6.

to provide desired testimony.[19] These documents also contain information on the effectiveness of various means to undermine inconvenient opinions from government agencies such as the Environmental Protection Agency. These pertain mainly to ways of undermining the scientific credibility of experts.

The negation of the climate change problem is just one example of the wider phenomenon known as the product defense industry or the "merchandizing of doubt and ignorance."[20] This involves experts, think tanks, scientific institutions, consulting firms and PR agencies who offer professional services aiming to reinforce public perception on the safety of given products or solutions. They create an atmosphere of controversy around expert opinions that could be harmful to certain stakeholders (like, tobacco fuel and biotechnology companies). Among those studying the on-demand production of controversy is the American science historian Robert N. Proctor, who was also one of the first to testify against the tobacco companies on the United States. He dubs his area of study "agnotology."[21] Becoming more widely used in reference to the trends described here are terms like denialism and negationism. In his aforementioned book *Storms of My Grandchildren: The Truth about the Coming Climate Catastrophe and our Last Chance to Save Humanity*, Hansen calls those inciting doubt "contrarians" – opponents of prevailing opinion, opponents of consensus.[22]

The denying of facts and fabricating doubt are rhetorical strategies very pervasive in the twenty-first century. Their use is political in that they often generate considerable delays, or even long-term stagnation in processes for the regulation of potentially harmful products, legal processes and political decisions. The tactics of the doubt production industry rely on proliferating contradictions and doubt on the national and international levels. These are institutionalized agendas by way of which controversy is professionally created by undermining

---

19 Oreskes, Conway, *Merchants of Doubt*, p. 6.
20 See David Michaels, *Doubt Is Their Product. How Industry's Assault on Science Threatens Your Health* (Oxford: Oxford University Press, 2008); Robert N. Proctor, *Cancer Wars. How Politics Shapes What We Know and Don't Know about Cancer* (New York: Basic Books, 1995); Brian Wynne, "When Doubt Becomes a Weapon," *Nature* 466/2010, pp. 441–442.
21 As it appears, to a certain extent, we can treat the manufacture of doubt as a subject of study in the field of risk management studies. It must be noted, however, that due to its nature, activity of this kind is not publicized by those involved in it.
22 James Hansen, *Storms of My Grandchildren. The Truth about the Coming Climate Catastrophy and Our Last Chance to Save Humanity* (New York: Bloomsbury Press, 2009), pp. 54–58.

the status of scientific facts which may be detrimental to particular parties. The existence of the phenomenon shows how a cacophony of opposition, even when failing to build a solid counterargument or counter-fact, can be effective in maintaining a sense of public controversy surrounding given information.

Hyperdenial Mechanics I: The Players

When it comes to the problem of climate change, the strongest players in the disinformation campaigns are mainly conservative scientific institutions and think tanks like the George C. Marshall Institute, the Tobacco Industry Research Committee, the Cato Institute, the Heartland Institute and the American Petroleum Institute (API).[23] These entities enjoyed extensive funding from tobacco companies and corporations like the ExxonMobil Corporation, the Philip Morris Corporation, the Rio Tinto Group and General Motors.[24] Some of them, as Klein states in her book, choose to remain non-transparent, not publishing the names of their sponsors. This is the case with the Heartland Institute, for example. Various quasi-scientific institutes founded by corporations, like the Tobacco Industry Research Committee, conducted their own, often inconclusive, research.

Of the 56 books voicing skepticism toward the hypothesis of global warming being anthropogenic published in the 1990s and studied by Peter J. Jacques, Riley E. Dunlap and Marc Freeman, no less than 92 % were written by authors with ties to American right-wing, conservative foundations sponsored by the tobacco

---

23 David Demeritt, "Science Studies, Climate Change and the Prospects for Constructivist Critique," *Economy and Society* 35(3)/2006: p. 454; Anthony Giddens, *The Politics of Climate Change*, p. 119.
24 Oreskes, Conway, *Merchants of Doubt*, p. 232. In 2006, the Royal Society issued a letter to ExxonMobil denouncing the practices of this concern and calling for it to desist in sponsoring initiatives aiming to breed confusion on climate change, Steven A. Kolmes, "Climate Change," p. 36. ExxonMobil was also criticized by the American Union of Concerned Scientists (UCS) for leading "the most sophisticated and most successful disinformation campaign since the tobacco industry misled the public" (the UCS is a non-profit organization established in the 1960s by MIT students, today comprising 400,000 citizens and scientists; see http://www.ucsusa.org/about/, accessed: 3.09.2021). In 2006, ExxonMobil declared it would no longer fund institutions which manufacture doubt surrounding climate change. However, in 2013 Greenpeace claimed that ExxonMobil did not fulfil this promise (http://www.greenpeace.org/usa/en/campaigns/global-warming-and-energy/stop-exxonmobil/, accessed: 10.10.2013 – website has been transformed).

and fuel industries. In the 1980s, thirteen such books had been published and all of them were connected to interest groups of the kind identified above.[25] The following decade brought a five-fold increase in the publication of books skeptical of the human cause of climate change.

In the United States, the most influential experts involved in the campaign to negate the anthropogenic causes of global warming were physicists and not climatologists. They did not conduct any meaningful research in the field of climatology.[26] Among the names appearing most often are Frederick Seitz, Robert Jastrow (an astrophysicist), William A. Nierenberg, and S. Fred Singer. This small group of retired natural science celebrities managed to stir up controversy in public opinion despite a prevailing consensus among scientific experts specializing in climate change.[27] These scientists had played important roles during the Cold War, enjoying considerable influence in the White House,[28] and all of them were affiliated with the conservative think tank the George C. Marshall

---

25 Peter J. Jacques,, Riley E. Dunlap, Marc Freeman, "The Organisation of Denial: Conservative Think Tanks and Environmental Scepticism," *Environmental Politics* 17(3)/2008: pp. 349–385; see also Riley E. Dunlap, Peter J. Jacques, "Climate Change Denial Books and Conservative Think Tanks. Exploring the Connection," *American Behavioral Scientist* 57(6)/2013: pp. 705–706; Oreskes, Conway, *Merchants of Doubt*, p. 253.
26 In his article, Leo Elshof discusses the activity of Timothy Ball, a Canadian skeptic of climate change being human-caused. Much like his American physicist counterparts, the geographer Ball did not publish any work in the field of climatology, Leo Elshof, "Changing Worldviews to Cope with a Changing Climate," in: *Climate Change and Philosophy. Transformational Possibilities*, ed. Ruth Irwin (New York: Continuum, 2010), p. 89.
27 David Demeritt, "Science Studies," p. 460.
28 Oreskes, Conway, *Merchants of Doubt*, p. 213. The concerns of the opponents of pre-environmental regulations (like fuel and climate taxes or pollution limits) related to loss of sovereignty and a fear of totalitarianism. They feared that free-market principles would be compromised by new central planning forms. In this, a significant role was surely played by ideological and personal calls to action. For example, Frederik Seitz was not only the president of the National Academy of Science and of Rockefeller University but also a co-author of research on the atomic bomb and a supporter of the Vietnam War. His patriotic rhetoric contributed to the development of arms of mass destruction. He espoused traditional American values like civil liberties, anti-communism, the free market and the individual's right to accumulate wealth.

Institute.[29] Seitz and Singer worked for the tobacco industry while also being involved in military research during the cold war (including work on the atomic bomb project).

Fred Singer, for example, had a part in the exploitation of the figure of Revelle, the aforementioned climate researcher who stressed the significance of the problem of climate change caused by human activity and greatly influenced the stance of Al Gore. After Revelle suffered a heart attack at the age of 81 in 1990, Singer approached him with the intention of producing a collaborative article on climate change, which the former was unable to decline outright. As the article was being edited, Singer ignored Revelle's remarks and in 1991 published a version of the article in the non-reviewed journal Cosmos which stated that the hypothesis of climate change is an unlikely one. During the 1992 presidential election campaign in which Gore ran, the media cited the article written by Singer and Revelle, stating that even Gore's mentor had reversed his opinion on the causes of climate change toward the end of his life. These events outraged the family and friends of the late Revelle.[30]

Frederick Seitz,[31] meanwhile, perpetrated a public attack on Benjamin D. Santer, a climatologist and co-author of the 1997 IPCC report. In response to his argument on the sun's supposed influence on global warming, Veerabhadran Ramanathan, an atmosphere researcher and climatologist, suggested analyzing the vertical temperature structure of Earth's climate. If the sun was responsible for global warming, Ramanathan argued, then a temperature increase would be noted in the upper stratosphere as well as in the lower troposphere. If, however, global warming, was caused by greenhouse gasses lingering in the lower layers of the atmosphere, the troposphere's temperature would increase but not the stratosphere's. One of the climatologists who conducted analyses and observations following Ramanathan's suggestion was Santer. His findings indicated that the cause of global warming could not be a change in the sun's activity because the temperature of the stratosphere had dropped. Subsequent to the findings of Santer and other researchers, the boundary of the stratosphere in relation to the troposphere was redefined as it is precisely temperature that determines the boundaries of these layers. Santer's extensive empirical response

---

29 The George C. Marshall Institute was founded in 1984. Since the mid-1990s it has been funded not only by private donors but also the conservative Olin Foundation and various corporations.
30 Oreskes, Conway, *Merchants of Doubt*, p. 190 ff.
31 Since 1979, Frederik Seitz has received 43.4 million dollars to finance scientific grants by the tobacco industry Oreskes, Conway, *Merchants of Doubt*, p. 29.

to Seitz's criticism (he, being a physicist, did not go into details on the climate) appeared in a much abridged version in *The Wall Street Journal*. That article did not name the 40 other scientists who co-authored the IPCC report and supported Santer's conclusion.[32]

One of the most high-profile figures to whom the global warming disinformation campaign is indebted is Steven Milloy, a Fox News pundit who would pin the label of "junk science" on scientific findings he did not agree with. In an effort to undermine certain scientific data, he presented them in a highly manipulated and unqualified manner as being controversial and methodologically unsound. It eventually came out that Milloy had ties to the Cato Institute and the Philip Morris Corporation, and that he enjoyed financial backing from the tobacco and oil industries (including ExxonMobil). Some of the other most active negators are identified by Naomi Klein. Among them are Marc Morano, the founder of the Climate Depot website, which professes skepticism toward climate destabilization, and Chris Horner of the Competitive Enterprise Institute, which harasses climatologists with burdensome lawsuits.[33]

As Charles N. Herrick and Dale Jamieson demonstrate in one of their articles, among the active players in the processes described here were also US mass media outlets, which (like Milloy) in 1995–2000 systematically threw around the term "junk science" even though empirical data on the environment exhibited no signs of epistemic frailty.[34] Herrick and Jamieson state that the term "junk science" served as nothing more than a powerful rhetorical tool for misdirection. It was used effectively in support of the ideology and practices of anti-regulation campaigns. Accused of being junk science were those studies which supported the need to impose regulations or called for public decisions on the environment or health. Narratives speaking of junk science typically appeared in articles which opposed regulation: 84 % of the articles examined were unequivocally against regulation.[35] Many of the articles criticizing regulation also expressed a longing for a return to "solid science," suggesting that scientifically sound practices had vanished.

Herrick and Jamieson argue that even though the methodological and procedural uselessness of certain research was not hard to prove (lack of bibliographic

---

32 Oreskes, Conway, *Merchants of Doubt*, p. 208.
33 Naomi Klein, *This Changes Everything*, p. 47.
34 Charles N. Herrick, Dale Jamieson, "Junk Science and Environmental Policy. Obscuring Public Debate with Misleading Discourse," *Philosophy & Public Policy Quarterly* 21(2–3)/2001, pp. 11–16.
35 Charles N. Herrick, Dale Jamieson, "Junk Science," p. 14.

references, publishing in non-peer-reviewed journals, the author's lack of qualifications in a given field, data manipulation), the cases they studied tended not to rely on credible evidence. Instead, they were dominated by accusations or hazy suggestions of detrimental economic or social consequences expected from regulation. Authors accused scientists of focusing excessively on scientific findings identifying risks that should be regulated and attacked them for allegedly ignoring contrary findings (which were rarely cited). They made baseless claims that research data was simply falsified or collected by scientists affiliated with entities poised to benefit from regulation.[36]

Analyzing the spread of specific controversies related to current knowledge on the climate, we can observe a clear dissymmetry in the resources employed by climatologists presenting their finding and the highly-organized interest groups which skillfully exploit the available tools of influence to shape public opinion. Particularly striking is data published by STS researchers on the financial support enjoyed by research institutes which undermine scientific findings that could prove detrimental to the interests of tobacco, fuel and arms concerns.[37]

---

36 Charles N. Herrick, Dale Jamieson, "Junk Science," p. 13. It is worth adding that due to the instrumental use of the term "junk science" in disinformation campaigns, Herrick and Jamieson suggest in their article that citizens should be educated on how complex science is and how it could be utilized in developing environmental policy.

37 In 2000–2005, ExxonMobil provided around 8 million dollars to select scientists "working on" the problem of global warming, Oreskes, Conway, *Merchants of Doubt*, p. 246. The data on the fuel industry's spending on campaigns aiming to manipulate public opinion regarding climate change is shocking, Leo Elshof, "Changing Worldviews," p. 95. Up to 2010, BP and ExxonMobil invested more in ad campaigns portraying them as pro-environmental companies than on any research into alternative energy sources, Leo Elshof, "Changing Worldviews," pp. 98–99. Used by the doubt-production industry is the Advancement of Sound Science Coalition (TASSC), an organization sponsored by the tobacco industry which, in the midst of the controversy surrounding the hazards of second-hand smoke, utilized the public media to influence three million readers, Oreskes, Conway, *Merchants of Doubt*, p. 51. It cannot be overstated how formidable these bodies are. Speth states that ExxonMobil has greater economic power that 180 of the world's countries, James Gustave Speth, *The Bridge at the Edge of the World. Capitalism, the Environment, and Crossing from Crisis to Sustainability* (New Haven, London: Yale University Press, 2008), p. 62.

Hyperdenial Mechanics II: Rhetorical Strategies

Characteristically, the controversy-on-demand industry wages war against science by exploiting science's rhetorical tools and turning its mechanisms against it.[38] It is possible to generate an atmosphere of controversy around specific complex scientific issues because the cornerstone of science's public image is its openness to criticism. This works well given the very nature of the problem: the vastness of the phenomenon, its level of complexity, the need for an interdisciplinary approach, methodological instability – all are exploitable in stirring up doubt. As mentioned earlier, the complexity of the climate change problem and the interdisciplinariness of climatology understandably elicit a certain amount of doubt (related to methodology, theory, and the even the competence of the scientists). There is interpretive flexibility regarding specific conclusions in the complex web of procedures that make it possible to assess the state of the climate.

The producers of doubt on demand put forth competing hypotheses (like, for example, the influence of prions, stress or genetic makeup on the likelihood of being diagnosed with lung cancer). They constantly emphasize medical science's or natural science's inability to provide definitive proof (using terms like "there is no proof"). They ignore available empirical evidence and inconvenient findings; hurl accusations at their opponents without providing evidence; use out-of-context quotes from their targets, and even often resort to surreptitiously quoting themselves, referencing their own opinions under the guise of citing other experts.[39]

An effective tactic used repeatedly in disinformation campaigns is the strategy of "paralysis by analysis,"[40] which involves constant insistence on the need to conduct further, more detailed study, demanding ever more precise information, exploring all related circumstances and dwelling on the complexity of the problem at hand.

Moreover, the institution of science, based on the mechanism of highly specialized articles undergoing review by peer experts before being published in specialty periodicals, occurs to be defenseless in the face of smear tactics aiming to undermine its credibility in the eyes of the public. In effect "unscientific claims were being circulated broadly, but the scientists' refutation of them was published where only fellow scientists would see it."[41] In the campaign

---

38 Oreskes, Conway, *Merchants of Doubt*, p. 13.
39 Oreskes, Conway, *Merchants of Doubt*, p. 166.
40 Oreskes, Conway, *Merchants of Doubt*, p. 148.
41 Oreskes, Conway, *Merchants of Doubt*, p. 195.

against global warming, it is mass media outlets like *The New York Times*, *The Washington Post* and *Newsweek* that played a crucial role. They eagerly published skeptical voices in their constant pursuit of sensation. They failed to recognize, or were not interested in recognizing, the fact that they were complicit in a deliberate and specially-funded campaign to spread disinformation. On their pages, they presented the arguments of both sides in equal proportions.[42]

The reason why media disinformation campaigns are so effective is not only that mainstream media in the USA, like *The New York Times* or *The Washington Post* are conservative. Playing an important role was the since-discredited doctrine of the media being impartial. Existing in American law since 1949, this rule dictated that both sides of a given controversy must be given equal representation. It was part of the Communications Act passed at that time, which concerned the publication of information of public significance. It must, however, be noted that in the case of a scientific consensus being undermined by organized interest groups fronted by a few science celebrities, such an imperative of equality becomes a bias of sorts. Logic states that the media, instead of presenting controversial issues symmetrically, ought to cover them proportionally (i.e. if 2 % of scientists are skeptical of a given hypothesis they should receive 2 % and not 50 % of the media's attention). Oreskes and Conway write that media representatives failed to understand that they are dealing with highly-organized and well-funded campaigns to spread disinformation and not with real controversy.[43]

As one analysis of the means of presenting the problem of global warming in the media discovered, in the USA in 1988–2002, 53 % of the articles studied took a balanced approach to the information, devoting equal coverage to skeptics and

---

[42] These media outlets were under great pressure – when they failed to abide by the aforementioned symmetrical and "fair" strategy, magazine publishers received "persuasive" letters, Oreskes, Conway, *Merchants of Doubt*, pp. 214–215. As Klein points out in her book, negators regularly intimidate media outlets. To this day, each mention of climate change elicits a wave of hateful online attacks on the author or publication as well as deluges of spam to the inboxes on specific employees, Naomi Klein, *This Changes Everything*. Marcin Popkiewicz, meanwhile, emphasizes that when CNN reported on climate change it generated an avalanche of protests from petroleum and automotive corporations, Marcin Popkiewicz, *Świat na rozdrożu*, second edition (Katowice: Wydawnictwo Sonia Draga, 2016), p. 428.

[43] Oreskes, Conway, *Merchants of Doubt*, p. 7; cf. Aaron M. McCright, Riley E. Dunlap, "Defeating Kyoto: The Conservative Movement's Impact on U.S. Climate Change Policy," *Social Problems* 50(3)/2003, p. 133; Colin Crouch, *Knowledge Corrupters: Hidden Consequences of the Financial Takeover of Public Life* (Cambridge, Malden: MA Polity Press, 2016), ch. 2.

scientists supporting the hypothesis of human-caused global warming. The 35 % of articles that focused on the stance of scientists arguing that climate change is human-caused still provided the skeptics with their due coverage. Meanwhile, a 2007 study of American radio coverage on climate change found that no less than 91 % of radio reports represented the skeptics' side.[44] The mechanism described here thus created an informational bias: though not controversial at all in the field of professional science, the media still presented the problem as a controversy.[45]

As it appears, in the case of the phenomenon of skepticism toward global warming in the United States, what we have seen is a kind of breakdown in scientific communication. Mainstream natural scientists were not overly interested in getting involved in public debate or in defending their findings against the skeptics' criticism. Also, the American media often denied them the opportunity to do so.

We may attempt to identify the possible reasons for the passivity exhibited by the academic community in the face the disinformation campaigns discussed here. Oreskes and Conway attribute this to something akin to alienation felt by experts (they did not understand the functioning mechanisms of mass media). Often, they did not know how to communicate their outrage effectively, limiting themselves to appearances in academic circles, specialty press, and public statements made through scientific associations.[46]

Many experts ostensibly assumed that the truth would come out on its own – that there was no need to waste time on public debate in its defense (this is evidently a repercussion of the philosophical stance of essentialism[47]). Also proving significant was that many experts feared sanctions from their peers if they took it upon themselves to speak in the name of their respective field. Also

---

44 John Halpin, *The Structural Imbalance of Political Talk Radio* (Washington: The Center for American Progress Free Press, 2007).

45 Maxwell T. Boykoff, Jules M. Boykoff, "Balance as Bias. Global Warming and the US Prestige Press," *Global Environmental Change* 14/2004, pp. 125–136; see also Maxwell T. Boykoff, *Who Speaks for the Climate? Making Sense of Media Reporting on Climate Change* (Cambridge, MA: Cambridge University Press, 2011).

46 Oreskes, Conway, *Merchants of Doubt*, pp. 262–265.

47 In line with this attitude, scientific activity serves mainly to discover new natural laws – to uncover facts. Once uncovered, these laws will defend themselves over time. Knowledge is an unproblematic good, and the pursuit of it is politically neutral or even axiologically redeeming in terms of human interests. Such a belief arises from the philosophy of the Enlightenment.

not insignificant were scientists' fears of being personally attacked in public and of possible costly legal battles against corporations or think tanks involved in the manufacture of doubt. It must be noted that a fear of legal action was not limited to scientists but also affected people in the media as well.[48]

Doubt-makers also attacked the authority of science at large, arguing that science has always been political, idealized, and instrumentalized. They had no qualms about undermining the worth of the Nobel Prize when it suited their purposes (when the attacks targeted Frank Sherwood Rowland, a chemist and ozone hole researcher who received the Nobel Prize with Paul Crutzen and Mario Molina in 1995). Seitz himself did not hold back from attacking the American National Academy of Sciences. The denialists clung to the strategy that science must be undercut by any means necessary of it took the side of regulation supporters.[49]

The specific rhetorical strategies applied by the movement to negate global warming include 1) rejecting the problem; portraying it as a fabrication of "ecoterrorists" or a tool used by atomic energy advocates in their political mission, 2) changing the subject; i.e. rejecting the hypothesis of climate change being human-caused and pointing to alternative explanations (like, for instance, the activity of the sun, Mars, volcanos, cosmic radiation, or urban development), 3) ignoring the fact that the risk lies not only in global warming but also in the systemic destabilization of the climate, in addition to stressing possibly beneficial consequences of global warming, 4) obscuring the subject by focusing on factors that are uncertain and not fully explainable, 5) asserting that the problem is unsolvable and discounting remedial measures as utopian, 6) relying on enlightened hopes rooted in the ideology of progress and the technological imperative, which state that future technological solutions will save us; that the future will fix itself whether we help it or not.[50] The Royal Society also published an analysis of the most popular arguments against the hypothesis of climate

---

48 Oreskes and Conway write that the Tobacco Industry Research Committee coerced journalists to report in an "unbiased" manner on research into the hazards of smoking.
49 Oreskes, Conway, *Merchants of Doubt*, pp. 64–65.
50 Similar tactics were used in earlier campaigns to manufacture doubt. For example, volcanic activity was used as an explanation for the phenomena of acid rain and the hole in the ozone layer. Interestingly, nearly all of the abovementioned rhetorical strategies appear in the controversial film which undermines the anthropogenic global theory, titled *The Great Global Warming Swindle*, directed by Martin Durkin and first broadcast in the UK on Channel 4 in 2007.

change being anthropogenic and of the rhetorical strategies most frequently employed by skeptics.[51]

One of the techniques used in campaigns to negate climate change is to create seemingly grass-roots or citizens' expert organizations. This was often accompanied by rhetorical tactics like giving pro-environmental-sounding names to organizations whose actual aim was to combat solutions intended for environmental protection. In their book *Betrayal of Science and Reason: How Antienvironmental Rhetoric Threatens Our Future*, Paul R. Ehrlich and Anne H. Ehrlich describe the conduct of such pseudo-environmental organizations as "aggressive mimicry."[52] Examples include the Canadian Coalition for Responsible Environmental Solutions, whose aim was to stifle initiatives for preventing global warming, and the Citizens' Alliance for Responsible Energy, which promoted the idea of petroleum drilling within Arctic natural reserves.[53] Other such organizations with neutral or even pro-environmental names include: the Friends of Science, the Natural Resources Stewardship Project, the International Climate Science Coalition (ICSC), IceCap, and the Science and Public Policy Institute (SPPI),[54] not to mention The Sahara Club or The National Wilderness Institute.[55]

Merchants of doubt also resorted to means that were utterly foreign to scientific circles. Oreskes and Conway cite the example of a petition, still circulating online since 1997, launched by Seitz (the so-called Oregon Petition), urging the US government to reject the anthropogenic global warming hypothesis. Seitz himself signed the petition as the former president of the National Academy of Sciences in an attempt to disguise it as being a legitimate initiative of scientific pedigree. The National Academy of Science was forced to call a special press conference in order to dissociate itself from Seitz's petition. But the damage had already been done (and the petition lives on, having garnered more than ten thousand signatures).

Fred Singer, on the other hand, started an institution to compete with the IPCC – the "non-IPCC" sponsored by the Heartland Institute. The controversy-making

---

51 The Royal Society, *Climate Change Controversies. A Simple Guide* (London: The Royal Society 2007). (21 October 2017), http://royalsociety.org/~/media/Royal_Society_Content/policy/publications/2007/8031.pdf.
52 Paul R. Ehrlich, Anne H. Ehrlich, *Betrayal of Science and Reason. How Antienvironmental Rhetoric Threatens Our Future* (New York: Island Press, 1998).
53 Hugh P. McDonald, "Denying Climate Change," p. 126.
54 Leo Elshof, "Changing Worldviews" p. 89.
55 Haydn W. Washington, *Human Dependence on Nature. How to Help Solve the Environmental Crisis* (New York: Earthscan/Routledge, 2013), p. 88.

industry also has not hesitated to create its own simulacra of scientific bodies, like tobacco magazines disguised as peer-reviewed science journals.[56]

Also quite interesting is the strategy of the well-known and widely-criticized book *The Skeptical Environmentalist* by Bjørn Lomborg,[57] who was affiliated with the Heartland Institute and the Hoover Institution. In this work, Lomborg attempts to change the subject; he ignores the gravity of the threat of climate destabilization by arguing that other global problems like hunger are today much more pertinent. The Danish author is also a technological optimist, an advocate of the technological imperative putting faith in the ability of future technologies to save us. Denmark's minister of science defended the publication against the ensuing backlash by emphasizing that it was not a scientific book, thereby suggesting that it could not be criticized for containing false theses, misleading information and problematic interpretations of statistical data.[58] Ultimately, in 2003, The Danish Committee for Scientific Dishonesty determined that Lomborg's book contained fabricated data.[59]

Not surprisingly, the institution attracting the bulk of the doubt-makers' onslaught is the IPCC.[60] The most common accusations against this institution cite its alleged bias and conflicts of interest (in 2010), its lack of a response to the accusations of conflict of interest (2011), its experts being unqualified,[61] and (to a lesser extent) errors in its reports. Also being challenged was the review

---

56 Oreskes, Conway, *Merchants of Doubt*, p. 144.
57 Bjørn Lomborg, *The Skeptical Environmentalist* (Cambridge: Cambridge University Press, 2001).
58 Oreskes, Conway, *Merchants of Doubt*, pp. 258–260.
59 Haydn W. Washington, *Human Dependence*, p. 89. It must be noted that Lomborg has since changed his stance on the problem of the climate catastrophe and has abandoned the skeptical point of view.
60 Cf. David Demeritt, "Science Studies" p. 454.
61 As Charles Krauthammer stated in The Washington Post, the environmental movement is simply socialism in disguise, Charles Krauthammer, "The New Socialism," *The Washington Post*, 2009 (December 2009 – actually the website is not active) http://www.washingtonpost.com/wpdyn/content/article/2009/12/10/AR2009121003163.html. Meanwhile as the aforementioned Pitman laments "I have been accused of being both a Marxist and a Fascist, of trying to destroy the Australian economy, trying to provide an avenue for mass migration, arguing for genocide (on the grounds that I have apparently argued for enforced population control – which I have not), supporting a single (presumably communist) world government, and a range of other increasingly bizarre accusations have been made," Steven Yearley, David Mercer, Andrew Pitman, Naomi Oreskes, Erik M. Conway, "Perspectives on Global Warming," p. 12.

procedure used in the IPCC reports. The reports tend to contain introductions that are controversial to political entities, their message often being disputed by governments.[62] Moreover, the IPCC being a large international organization, these is a shortage of independent experts to review their reports. It is also hard to argue with the statement that it is in the IPCC's structural interest not to question the validity of climate change or to even embellish the threats related to climate destabilization. STS researchers also point out that to assess our knowledge on the state of the climate is an endeavor of an entirely new kind. In an institution as complex as the IPCC we encounter completely novel mechanisms for negotiating the interpretation of research data and reaching consensus on specific points.[63]

In the article "Defeating Kyoto: The Conservative Movement's Impact on U.S. Climate Change Policy,"[64] Aaron M. McCright and Riley E. Dunlap outline how in 1990–1997 the conservative movement in the United States deceived the public on global warming. The authors define the conservative movement as a network, strongly influenced by the nation's elite, of private foundations,

---

62 It comes as no surprise that during their acceptance process, the reports of the IPCC meet with opposition from delegates of oil-producing countries like Saudi Arabia and Kuwait (that was the case in 1995, for instance). The panel's reports are also criticized as narratives which unjustly favor discourse and economic categories. One of the controversies, for example, concerned the means of estimating the economic value of citizens' lives in a given country. In the reports' analyses of social impact two types of responses to climate change are considered: emission reduction and adaptation. Both options are costly. The IPCC estimated the value of the average East Asian citizen's life as 1/15 that of a developed country's citizen (sic!). Pursuant to such a valuation, it was calculated that adaptation would be cheaper as it would apply to developing countries where the costs in terms of the value of citizens' lives would be lower than the economic costs of implementing greenhouse gas emissions limits, Steven Yearley, "Nature and the Environment in Science and Technology Studies," in: *The Handbook of Science and Technology Studies*, eds. Edward J. Hackett, Olga Amsterdamska, Michael Lynch and Judy Wajcman (Cambridge, MA: MIT Press, 2008), pp. 926–929. It must be added that the term "cost per life saved" (CPLS) was used earlier in international humanitarian programs in the 1970s (cf. Małgorzata Kaniewska, *Etyczno-prawne aspekty zmian klimatycznych* (Józefów: Wydawnictwo Wyższej Szkoły Gospodarki Euroregionalnej im. Alcide de Gasperi, 2017), p. 246.
63 Paul N. Edwards, Stephen H. Schneider, "Self-Governance and Peer Review in Science-for-Policy. The Case of the IPCC Second Assessment Report," in: *Changing the Atmosphere. Expert Knowledge and Environmental Governance*, eds. A. Miller Clark and Paul N. Edwards (Cambridge, MA: MIT Press, 2001), pp. 219–246.
64 McCright, Dunlap, "Defeating Kyoto."

think tanks, activists and entities espousing free-market beliefs and traditional values.[65] They stress that conservative views often strongly conflict with pro-environmental ideas, which typically go hand in hand with support for stronger regulation and active governmental policies.[66]

In July 1997, the Bill Clinton administration was forced to withdraw from the push to ratify the Kyoto Protocol. In 2001, George W. Bush declared that the USA had not committed to reducing emissions and did not intend to abide by the protocol. McCright and Dunlap believe this was a product of several overlapping strategies. The first involved reclassifying and redefining the global warming problem as a phenomenon that was insufficiently documented by science and as a process entailing beneficial and not catastrophic changes. At the same time, the conservatives portrayed any potential climate policy as a collection of projects that would do more harm than good, especially in terms of economic growth, the free market and national sovereignty.[67]

The second strategy relied on a deployment of resources through the use of think tanks affiliated with the conservatives. McCright and Dunlap analyzed the role of fourteen of the most influential conservative think tanks in the United States, like the John M. Olin Foundation, the Heritage Foundation, the Marshall Institute, the American Enterprise Institute, the Citizens for a Sound Economy Foundation, and the Cato Institute, as well as of various experts voicing skepticism on global warming (chief among them the astrophysicist Sallie Baliunas, Robert Balling Jr., the atmosphere researcher Richard Lindzen, Patrick Michaels and the environmental scientist S. Fred Singer). The authors found that a similar strategic approach had been employed in various earlier political campaigns initiated by the conservatives, including ones on immigration reform, education, social policy and affirmative action. In the United States, conservative think tanks function to stimulate social movements and to influence public opinion, doing so mainly thanks to substantial financial support from private individuals and some of the country's largest businesses.[68] In 1990–1997, these think tanks

---

65 McCright, Dunlap, "Defeating Kyoto," fn. 6.
66 McCright, Dunlap, "Defeating Kyoto," p. 353. Interestingly, after the ratification of the Kyoto Protocol large corporations like DuPont and BP (still in 1997), Shell, General Motor and Sun Oil (in 1998) publicly acknowledged the existence of the anthropogenic global warming problem, while the conservative movement in the USA continued to issue objections and doubts regarding this fact, McCright, Dunlap, "Defeating Kyoto," p. 369.
67 McCright, Dunlap, "Defeating Kyoto," p. 354.
68 McCright, Dunlap, "Defeating Kyoto," p. 356.

published and distributed 224 documents dealing with climate change. Among them were editorials in magazines, newspaper articles, books and research reports. These publications were disseminated through channels outside of traditional natural sciences communication. They concertedly and systematically cited skeptics who do not represent the mainstream of climate research. In the tense year of 1997, conservative organizations also ran radio and television spots and lent financial support to documentary television series critical of regulation and environmental policies, ones like *Against Nature, 700 Club* and *A Firing Line Debate* broadcast by PBS.

The third strategy described by McCright and Dunlap relies on lobbying and on a system of sponsored consultations, discussion panels, public speeches and press conferences in addition to activity aiming to generate publicity in the media. McCright and Dunlap compared the media visibility of the above-mentioned skeptics affiliated with conservative institutions versus the visibility of the scientific elite represented by renowned scientists publishing in respected peer-reviewed journals – people like Stephen Schneider, Frank Sherwood Rowland, Bert Bolin, James Hansen and Benjamin Santer. McCright and Dunlap scoured online archives to analyze a range of widely-read publications from the United States, including The Wall Street Journal, USA Today, The New York Times, The Los Angeles Times, The Washington Post, Chicago Tribune, and Newsday. In articles dealing with climate destabilization published in 1995 and 1996, the skeptics were cited more often than their counterparts from the scientific elite and in 1997 both sides were cited equally often.[69] The disinformation campaign also included personal attacks and insinuations of alleged scientific dishonesty against climatologists publishing in peer-reviewed journals who supported the theory of climate change being human-caused. They were accused of procuring government grants unfairly, with the conservative think tanks appointing special ethics committees to investigate potential violations.

The fourth strategy involved calling experts to testify to the US Congress (in which the Republican Party had won a 52.9 % majority in 1994) and submit to review by Congress commissions. In 1995, the US Congress held three inquiries into unethical conduct by scientists studying the hole in the ozone layer, climate destabilization and dioxins.[70] McCright and Dunlap reviewed 37 Congressional hearings concerning global warming in 1990–1997, in which a total of 287 statements were given. Playing a key role was the House Committee

---

69 McCright, Dunlap, "Defeating Kyoto," pp. 356, 365.
70 McCright, Dunlap, "Defeating Kyoto" pp. 358, 361.

on Science as it challenged the value of the existing environmental programs and regulations. The authors recall one particular hearing from 1995 in which a politician lambasted scientists' insistence on the importance of peer review as nonsense and pulling the wool over people's eyes! In 1992, 47.9 % of Congress hearings involved testimony on climate change from natural science experts. By 1997, that figure had dropped to just 3.8 %. At the same time, the percentage of hearings involving testimony from representatives of industry had risen to 53.8 % by 1997. Over that entire span, the participation of environmental organization representatives was marginal, constituting just 8.4 % of the hearings. The statistical data presented by McCright and Dunlap also shows a consistent rise in the number of skeptical "experts" affiliated with right-wing think tanks appearing before Congress. Whereas in 1992, 34 science experts and two skeptics gave testimony, in 1995 and 1997 it was one of each, and in 1996, it was four skeptics and six science experts.[71]

According to McCright and Dunlap, public indifference to the climate problem is due to the fact that the issue is too complex and perceived as too distant in time, one whose potential solution is considered extraordinarily costly. They believe this apathy to be the work of the denialists and of effective campaigns run by conservative think tanks. Public opinion in the United States has deemed climate destabilization to be an issue which does not necessitate political action.[72]

How to Preserve the Authority of Science?

It cannot be denied that the parties responsible for manufacturing skepticism on climate change have had spectacular success. The George C. Marshall Institute exerted heavy influence on the United States government in the 1990s, which led to the United States' abandoning the idea of raising fuel taxation (against the recommendation of the first IPCC report). Meanwhile, the contrived Climategate affair[73] preceding the 2009 UN Climate Change Conference in Copenhagen

---

71 McCright, Dunlap, "Defeating Kyoto," pp. 362–364.
72 McCright, Dunlap, "Defeating Kyoto," p. 367.
73 The Glaciergate and Climategate affairs related to the activity of the IPCC. Glaciergate concerned errors in the 2007 IPCC report. In 2009, the institution admitted that the section on melting glaciers contained an error in the forecasted date when certain glaciers in the Himalayas will melt. Specifically, the year 2035 was identified instead of 2350. Also, the report provided some data that was not reviewed in scientific literature. In 2010, the IPCC issued a statement that one paragraph of the 2007 report contains some scientific date that is may not be credible. The Climategate affair concerned

played a part in the diplomatic failure of that event, in which no concrete climate change response program was devised.

The damage done by the skepticism campaigns is far-reaching and very detrimental to the mechanisms enabling the circulation of scientific knowledge in that scientists have become very inhibited while discussing climate and environmental threats in fear of attacks and accusations in the mainstream media.[74]

The "merchants of doubt" phenomenon is evidence that in the risk society the role of the independent science expert has changed considerably. Now, in addition to experts we have "professionals" stirring up doubt and uncertainty. It has become clear that doubt can be created and bought, and that controversy is a marketable commodity. However, we must not forget that IPCC experts and pro-environmental scientists are subject to the same conditions as conservative experts. Any expert can have a conflict of interest or political bias. In one way or another, science is always a part of the network of interests, and all scientific findings can be interpreted in different ways. It is not hard to imagine, for instance, that parties with an interest in the continued production of existing goods and solutions will be keen to point out the dangers of new alternatives in an effort to delay their implementation.

Yet, we must not condition our political decisions on a hope of achieving absolute expert certainty. Nature is not objective, knowledge is not unproblematic, and there is no universal, humanity-wide conception of the common good. The controversies unfolding in front of our eyes show that paternalistic ideas of scientific certainty must be challenged (rationally) for being unrealistic. As evidenced by the history of climate change skepticism, the rhetoric of scientific

---

correspondence between the Climate Research Unit (CRU) and the University of East Anglia. In 2009 an anonymous individual leaked emails sent by CRU scientists. In one of the emails, the institution's director, the climatologist Philip D. Jones, stated that he had "completed Mike's [Michael Mann's] Nature trick" to "hide the decline" in temperature over the last decades of the 20th century. Jones was subsequently accused of falsifying research results. The climatologist explained that the wording in some of his emails was ill-advised while also emphasizing that data published by the CRU on climate change was based on a range of sources and not just on the specific part of the research he referred to in his email. The charges against the CRU were dropped in 2010. A British Parliamentary Committee, a Royal Society committee and a University of East Anglia committee (among others) did not find any malfeasance in the institution's operation. Jones was reinstated. Today, this affair is considered fraudulent, see Naomi Klein, This Changes Everything, p. 59.

74 Elshof, "Changing Worldviews" p. 93.

certainty validated not only the status of the IPCC's experts but also that of the doubt-manufacturing industry.[75]

Bruno Latour stresses that it is misleading to think of science as separate from politics. As long as we cling to the premise that science simply presents objective, indisputable facts and that politics will automatically fall in line with expert scientific knowledge, we will remain helpless (as we have been so far) against the disinformation and lobbying funded by the deep pockets of the fuel and automotive industries.

Science is by nature susceptible to controversy, doubt and revision. Even in laboratory work, all we really have are solid, interactive stability between theories and materials and empirical data collected with highly-calibrated apparatus.[76] Only when we admit that science, climate studies included, resembles politics, can we expect to see positive results. The co-creator of Actor-Network Theory says this: "When you meet climasceptics who have the nerve to call the IPCC "a lobby," it would be much more powerful to answer: "Of course it is a lobby, now let us see how many are you; where does your money come from. … in what world do you live, where, with what resources, for how long, what future do you envision for your kids, what sort of education do you wish to give them, in which landscape do you wish them to live.""[77]

An STS analysis of scientific controversy cases revealed that judgement on a given fact's solidity is always contingent on recognizing the institutional agendas in the professional, scientific support for that fact (which, in turn, makes it possible to assess the credibility of the experts weighing in). The only thing we have to rely on are the anonymous mechanisms of review by fellow scientists in professional literature and within institutions responsible for grant disbursal. We may acknowledge these processes as a crucial condition behind the norm of organized skepticism characteristic of the science ethos as understood by Robert Merton (regardless of how wishful and normative the American sociologist's stance was). These mechanisms guarantee neither Truth nor Objectivity. It would be preferable if public opinion could appreciate the conditions for professionalism in the

---

75 See Yearley, Mercer, Pitman, Oreskes, Conway, "Perspectives on Global Warming," p. 7.
76 On the subject of solid, interactive stability, see Ewa Bińczyk, *Technonauka w społeczeństwie ryzyka. Filozofia wobec niepożądanych następstw praktycznego sukcesu nauki* (Toruń: Wydawnictwo Naukowe UMK, 2012), p. 106 ff.
77 Bruno Latour, "Telling Friends from Foes in the Time of the Anthropocene," in: *The Anthropocene and the Global Environmental Crisis. Rethinking Modernity in a New Epoch*, eds. Clive Hamilton, Christophe Bonneuil, François Gemenne (London, New York: Routledge, 2015), pp. 151–152.

area of science and, above all, the importance of the review mechanisms. For that to happen, however, public opinion would have to have at its disposal reliable information from the mass media, which, in a transparent and democratic society is responsible for exposing scientific conflicts of interest. The controversial role of scientists sounding off on subjects outside of their field of expertise and of those whose earlier work has been challenged by the scientific community must be exposed.[78]

In this context, science studies upholding the slogans of socially constructed knowledge have met with accusations of being politically harmful and irresponsible.[79] We must, however, bear in mind that theses on socially constructed knowledge come in various forms[80] and that the constructivist desacralization of science's epistemically privileged position was often motivated by the necessity to ensure experts' public responsibility in a risk society.

In line with the postulates of Harry M. Collins and Robert Evans on the acknowledgement of a third wave of science study, we must carefully categorize the various kinds of expert opinions and lay knowledge.[81] Expert opinions have various sides – Nobel laureates in physics are not necessarily credible experts when it comes to the climate. Because of this, it is worthwhile to once again underscore that the voices of skeptics in the global warming debate are not subject to peer review procedures. Skeptics do most of their work online and not in reviewed literature.[82]

Oreskes and Conway also firmly defend science's authority, the foundations of which are rooted in institutional norms. In a risk society, in which scientific controversies are politically pitted against each other, we need more, not less, well-placed trust in experts. Though science does not guarantee access to the truth, it does provide useful footing on which we can structure our strategy for action. Science "only provides the consensus of experts, based on the organized accumulation and scrutiny of evidence."[83] We are able to distinguish scientific

---

78 Harry M. Collins, *Are We All Scientific Experts Now?* (Cambridge, Malden: Polity Press, 2014), pp. 103–114.
79 David Demeritt, "Science Studies," pp. 456–457.
80 Cf. Ian Hacking, *The Social Construction of What?* (Cambridge, London: Harvard University Press, 1999).
81 Harry M. Collins,, Robert Evans, "The Third Wave of Science Studies. Studies of Expertise and Experience," *Social Studies of Science* 32/2002, pp. 235–296; Harry M. Collins, *Are We All Scientific Experts Now?*.
82 Leo Elshof, "Changing Worldviews," p. 88; David Demeritt, "Science Studies," p. 473.
83 Oreskes, Conway, *Merchants of Doubt*, p. 268.

theories from misleading disinformation campaigns. But, we must understand the significance of institutional foundations guaranteeing cognitive success in the area of science. Chief among these are scrupulous review mechanisms as well as institutional criteria for fact-based, critical discussion in journals and scientific conferences. It is, above all, the mechanisms of anonymous peer review of scientific findings which guarantees that science is science and that it is something other than mere opinion.

As we have discovered though, underscoring consensus among scientists does not automatically lead to political action for the sake of climate protection. This is not an effective strategy, a fact which had already been identified by science and technology studies. Results of studies quantifying climatological consensus on the theory of climate change being human-caused began to surface around 2004. Having the greatest public resonance was an article by John Cook et al., which estimated that in the field of climatology, 97.1 % of articles openly state or at least accept without reservation the notion that climate change is caused by human activity.[84] From this it appears that undisputed scientific knowledge failed to sway public opinion in the climate destabilization controversy. We know today that the participants in and observers of controversy are more influenced by political values and ideologies than by the facts.[85]

According to science sociologists like Bruno Latour, we will not overcome the (epistemic) stalemates generated by the skepticism-makers by expecting a clear division to arise between the sphere of facts and the sphere of opinion.[86] The climate change problem sensitizes us to the fact that, in the twenty-first century scientific instruments, statistical models, institutions, political opinions, axiological truths, and hopes will "mingle" ever more closely. One of the main conclusions of the analysis in this article is that it is hard to identify objective facts and also that we cannot afford to take a subjective point of view. In a risk society, the credibility of expert opinions is a commodity that will remain in the balance of continued political debate. Like it or not, in the context of the global warming

---

84 John Cook et al. "Quantifying the Consensus on Anthropogenic Global Warming in the Scientific Literature." *Environmental Research Letters* 8/2013, (15 March 2018) http://iopscience.iop.org/article/10.1088/1748-9326/8/2/024024/pdf; cf. Warren Pearce, Reiner Grundmann, Mike Hulme et al., "Beyond Counting Climate Consensus," *Environmental Communication* 11(6)/2017: pp. 723–730.

85 Warren Pearce, Reiner Grundmann, Mike Hulme et al., "Beyond Counting."

86 Bruno Latour, "Waiting for Gaia. Composing the Common World through Arts and Politics," lecture at The French Institute, London, November 2011 (5 March 2013) http://www.bruno-latour.fr/node/446, p. 7.

problem we are conducting experimental politics of nature as we try to build a harmonious collective future on the basis of this "flexible" data we possess on the state of the climate. In acknowledging the importance and objectivity of individual scientific problems, what we are really doing is deciding about what kind of future we envision for future generations and who will be our allies in the plans for creating that future.

Bibliography

Bińczyk, Ewa. *Technonauka w społeczeństwie ryzyka. Filozofia wobec niepożądanych następstw praktycznego sukcesu nauki.* Toruń: Wydawnictwo Naukowe UMK, 2012.

Boykoff, Maxwell T. *Who Speaks for the Climate? Making Sense of Media Reporting on Climate Change.* Cambridge, MA: Cambridge University Press, 2011.

Boykoff, Maxwell T., Jules M. Boykoff. "Balance as Bias. Global Warming and the US Prestige Press." *Global Environmental Change* 14/2004, pp. 125–136.

Carson, Rachel. *Silent Spring.* Boston, MA: Houghton Mifflin, 1962.

Collins, Harry M. *Are We All Scientific Experts Now?* Cambridge, Malden: Polity Press, 2014.

Collins, Harry M., Robert Evans, "The Third Wave of Science Studies. Studies of Expertise and Experience." *Social Studies of Science* 32/2002, pp. 235–296.

Cook, John et al. "Quantifying the Consensus on Anthropogenic Global Warming in the Scientific Literature." *Environmental Research Letters* 8/2013, (15 March 2018) http://iopscience.iop.org/article/10.1088/1748-9326/8/2/024024/pdf.

Crouch, Colin. *Knowledge Corrupters. Hidden Consequences of the Financial Takeover of Public Life.* Cambridge, Malden: MA Polity Press, 2016.

Demeritt, David. "Science Studies, Climate Change and the Prospects for Constructivist Critique." *Economy and Society* 35(3)/2006, pp. 453–479.

Dunlap, Riley E., Peter J. Jacques. Climate Change Denial Books and Conservative Think Tanks. Exploring the Connection. *American Behavioral Scientist* 57(6)/2013, pp. 705–706.

Edwards, Paul N., Stephen H. Schneider. "Self-Governance and Peer Review in Science-for-Policy. The Case of the IPCC Second Assessment Report," in: *Changing the Atmosphere. Expert Knowledge and Environmental Governance*, eds. A. Miller Clark and Paul N. Edwards, (Cambridge, MA: MIT Press, 2001), pp. 219–246.

Ehrlich, Paul R., Anne H. Ehrlich. *Betrayal of Science and Reason. How Antienvironmental Rhetoric Threatens Our Future.* New York: Island Press, 1998.

Elshof, Leo. "Changing Worldviews to Cope with a Changing Climate," in: *Climate Change and Philosophy. Transformational Possibilities,* ed. Ruth Irwin (New York: Continuum, 2010), pp. 75–108.

Giddens, Anthony. *The Politics of Climate Change.* Malden, MA: Polity Press, 2009.

Hacking, Ian. *The Social Construction of What?* Cambridge, London: Harvard University Press, 1999.

Halpin, John. *The Structural Imbalance of Political Talk Radio.* Washington: The Center for American Progress, Free Press, 2007.

Hamilton, Clive. "Theories of Climate Change." *Australian Journal of Political Science* 47(4)/2012, pp. 721–729.

Hansen, James. *Storms of My Grandchildren. The Truth about the Coming Climate Catastrophy and Our Last Chance to Save Humanity.* New York: Bloomsbury Press, 2009.

Herrick, Charles N., Dale Jamieson. "Junk Science and Environmental Policy. Obscuring Public Debate with Misleading Discourse." *Philosophy & Public Policy Quarterly* 21(2–3)/2001, pp. 11–16.

Jacques, Peter J., Riley E. Dunlap, Marc Freeman. "The Organisation of Denial. Conservative Think Tanks and Environmental Scepticism." *Environmental Politics* 17(3)/2008, pp. 349–385.

Kaniewska, Małgorzata. *Etyczno-prawne aspekty zmian klimatycznych.* Józefów: Wydawnictwo Wyższej Szkoły Gospodarki Euroregionalnej im. Alcide de Gasperi, 2017.

Klein, Naomi. *This Changes Everything. Capitalism vs. The Climate.* New York: Simon & Schuster, 2014.

Kolmes, Steven A. "Climate Change. A Disinformation Campaign." *Environment. Science and Policy for Sustainable Development* 53(4)/2011, pp. 33–37.

Krauthammer, Charles. "The New Socialism." *The Washington Post,* 2009. (11 December 2009 – actually website in not active). http://www.washingtonpost.com/wpdyn/content/article/2009/12/10/AR2009121003163.html.

Latour, Bruno. "Waiting for Gaia. Composing the Common World through Arts and Politics," lecture at The French Institute, London, November 2011. (5 March 2013) http.://www.bruno-latour.fr/node/446.

Latour, Bruno. "Telling Friends from Foes in the Time of the Anthropocene," in: *The Anthropocene and the Global Environmental Crisis. Rethinking*

Modernity in a New Epoch, eds. Clive Hamilton, Christophe Bonneuil, François Gemenne. London, New York: Routledge, 2015, pp. 145–155.

Lomborg, Bjørn. *The Skeptical Environmentalist*. Cambridge: Cambridge University Press, 2001.

McCright, Aaron M., Riley E. Dunlap. "Defeating Kyoto. The Conservative Movement's Impact on U.S. Climate Change Policy." *Social Problems* 50(3)/2003, pp. 348–373.

McDonald, Hugh P. "Denying Climate Change – Still!" *Capitalism, Nature, Socialism* 22(2)/2011, pp. 125–128.

Michaels, David. *Doubt Is Their Product. How Industry's Assault on Science Threatens Your Health*. Oxford: Oxford University Press, 2008.

Oreskes, Naomi, Erik M. Conway. *Merchants of Doubt. How a Handful of Scientists Obscured the Truth on Issues from Tobacco Smoke to Global Warming*. New York: Bloomsbury Press, 2010.

Pearce, Warren, Reiner Grundmann, Mike Hulme et al. "Beyond Counting Climate Consensus." *Environmental Communication* 11(6)/2017, pp. 723–730.

Popkiewicz, Marcin. *Świat na rozdrożu*. Second edition. Katowice: Wydawnictwo Sonia Draga, 2016.

Proctor, Robert N. *Cancer Wars. How Politics Shapes What We Know and Don't Know about Cancer*. New York: Basic Books, 1995.

Rasmussen Reports. "44% Say Global Warming Due To Planetary Trends, Not People." Research conducted by Rasmussen LLC on the 19th January 2009. (10 October 2013 – currently website is not active) http://www.rasmussenreports.com.

Rasmussen Reports. "Energy Update." 2012. (10 October 2013 – currently website is not active) http://www.rasmussenreports.com/public_content/politics/current_events/environment_energy/energy_update.

Ripple, William J., Christopher Wolf, Thomas M. Newsome et al. "World Scientists' Warning to Humanity. A Second Notice." *BioScience* 67(12)/2017, pp. 1026–1028.

Speth, James Gustave. *The Bridge at the Edge of the World. Capitalism, the Environment, and Crossing from Crisis to Sustainability*. New Haven, London: Yale University Press, 2008.

Stern, Nicholas. *The Global Deal. Climate Change and the Creation of a New Era of Progress and Prosperity*. New York: Public Affairs, 2009.

The Pew Research Center. "Fewer Americans See Solid Evidence on Global Warming," 2009. (2 November 2015) http://pewresearch.org./pubs/1386/cap-and-trade-global-warming-opinion.

The Pew Research Center. "Modest Rise in Number Saying There is 'Solid Evidence' of Global Warming," 2011. (2 November 2015) http://www.people-press.org/files/legacy-pdf/12-1-11%20Global%20warming%20release.pdf.

The Royal Society. "Climate Change Controversies. A Simple Guide." London: The Royal Society, 2007. (21 October 2017) http://royalsociety.org/~/media/Royal_Society_Content/policy/publications/2007/8031.pdf.

Washington, Haydn W. *Human Dependence on Nature. How to Help Solve the Environmental Crisis.* New York: Earthscan/Routledge, 2013.

Washington, Haydn W., John Cook. *Climate Change Denial. Heads in the Sand.* London: Earthscan Press, 2011.

Wynne, Brian. "When Doubt Becomes a Weapon." *Nature* 466/2010, pp. 441–442.

Yearley, Steven. "Nature and the Environment in Science and Technology Studies," in: *The Handbook of Science and Technology Studies*, eds. Edward J. Hackett, Olga Amsterdamska, Michael Lynch, Judy Wajcman. Cambridge, MA: MIT Press, 2008, pp. 921–947.

Yearley, Steven, David Mercer, Andrew Pitman, Naomi Oreskes, Erik M. Conway. "Perspectives on Global Warming. Book Symposium." *Metascience* 21(1)/2012, pp. 1–29.

Michał Wróblewski, Wojciech Goszczyński

Polish Smog: Metrological Controversies and Conflicting Ontologies[1]

Abstract: Since 2015 the problem of smog is the subject of lively public debate in Poland. Smog reports appear in the public transport of the largest cities, on the radio or on the Internet portals. Cheap air quality measurement devices and applications have appeared on the market, providing data on the current state of pollution, from which data are communicated via social media. The problem related to obtaining data, measurement methodologies or infrastructure related to air quality monitoring has with time become as controversial as smog itself. In the text we focus on air quality monitoring infrastructure and controversies related to smog data collection standards and infrastructures. We are interested in: 1) relationships between different actors dealing with smog and their infrastructures; 2) relationships between metrologies and ontology of the objects they refer to. We mainly look at the dynamics and the subject of conflicts associated with measurement and see how different regimes and practices enact different smog ontology. We first reconstruct the map of actors involved in the controversy, then describe the conflicts around air quality monitoring, and, point to the ontological dimension of the controversy.

Keywords: environmental conflicts, air quality monitoring, information infrastructures, ontologies, metrological chains

Introduction

Poland as a communist country became a subject of strong and violent industrialization after World War II. Until 1989, heavy industry played a leading role in the Polish economy and contributed to the increase in the level of pollution. The situation began to change after the political transformation. As part of the economic reforms, many industrial plants collapsed and those that survived the neoliberal shock therapy had to adapt to European regulations through installing of purification equipment. However, air pollution resulting from the so-called low emission caused by burning poor quality coal in household old

---

1   This research was funded by Polish National Science Centre, grant number UMO-2017/25/Z/HS6/03046.

furnaces[2] remained a problem. This concerns mainly pollutants related to PM10 and PM2.5 particulate matter and benzo(a)pyrene. The problem is so serious that, according to official data,[3] the population of Poland is practically breathing the worst air in the European Union.

The problem of smog became the subject of lively public debate a few years ago. In 2015 and 2016, several smog episodes were recorded in various regions of Poland, including the capital city of Warsaw.[4] The national media at that time began to increasingly report in the alarming tone the high concentrations of particulate matter. Smog reports now appear in the public transport of the largest cities, on the radio or on the Internet portals. A specific tracking culture has developed around the smog phenomenon.[5] Cheap air quality measurement devices and applications have appeared on the market, providing data on the current state of pollution, from which data are communicated via social media. The problem related to obtaining data, measurement methodologies or infrastructure related to air quality monitoring has with time become as inflammatory as smog itself. As we will try to show in this article, various actors in Poland argue over how to measure pollution, what equipment to use and how to present data to the public.

Given the growing importance of environmental problems in the public debate, pollution data are an interesting subject for research.[6] Data production is related to specific infrastructures that are not simply transmitters of knowledge about the outside world, but co-create their object of reference – they organize the work of experts and influence how a specific phenomenon exists in society.[7] From the ANT perspective, air is a hybrid, ontologically heterogeneous phenomenon that exists in networks of relationships consisting of human

---

2   GIOŚ (Główny Inspektorat Ochrony Środowiska). *Pyły Drobne w Atmosferze. Kompendium wiedzy o zanieczyszczeniu powietrza pyłem zawieszonym w Polsce* (Warszawa: Główny Inspektorat Ochrony Środowiska, 2016).
3   EEA (European Environmental Agency). *Air Quality in Europe – 2018 Report* (Luxembourg: Publications Office of the European Union 2018), pp. 26–34.
4   GIOŚ (Główny Inspektorat Ochrony Środowiska). *Analiza Wybranych Epizodów Wysokich Stężeń pyłu PM10 z lat 2013–2016* (Warszawa: Główny Inspektorat Ochrony Środowiska, 2017).
5   Deborah Lupton, *The Quantifed Self. Cambridge*, (Cambridge: Polity Press 2016).
6   Jennifer Gabrys, Helen Pritchard and Benjamin Barratt, "Just Good Enough Data: Figuring Data Citizenships Through Air Pollution Sensing and Data Stories," *Big Data & Society* 3(2)/2016, pp. 1–14.
7   Geoffrey Bowker, and Susan L. Star, *Sorting Things Out: Classifications and Its Consequences*. (Cambridge: MIT Press 1999).

and non-human, social and material factors.[8] It is also an example of a phenomenon that exists through the existence of metrological chains,[9] which include measurement standards and methodologies or measuring devices. In the case of pollution, there are different metrological, which means that in respect of smog, multiple ontologies[10] and ontological conflicts appear. This is associated with the fact that the production of environmental data is a process related to uncertainty and discontinuity[11] and an area where different epistemic cultures collide.[12] Epidemiologists and mathematical modelling specialists are a good example of varying approaches to air pollution and health impacts as they employ different methodologies and infrastructures.[13] Differences can also be seen in the relations between public institutions and citizens involved in environmental movements. The work of activists is often described as an example of citizen science[14] or embodied social movements.[15] Although quantitative indicators are a common denominator for experts and citizens, they adopt different strategies for handling pollution data.[16] Experts and citizens may have a different approach to the time of exposure to harmful pollution, whereby the definition of harm as well

---

8   Emma Garnett, "Developing a Feeling for Error: Practices of Monitoring and Modelling Air Pollution Data," *Big Data & Society* 3(2)/2016, pp. 1–12.
9   Bruno Latour, *Science in Action. How to Follow Scientists and Engineers Through Society* (Cambridge Massachusetts: Harvard University Press, 1987).
10 Annemarie Mol, *Body Multiple: Ontology in Medical Practice*, (Durham-London: Duke University Press, 2002).
11 Ingmar Lippert, "Failing the Market, Failing Deliberative Democracy: How Scaling Up Corporate Carbon Reporting Proliferates Information Asymmetries," *Big Data & Society* 3(2)/2016, pp. 1–13.
12 Karin Knorr-Cetina, *Epistemic Cultures. How the Sciences Make Knowledge* (Cambridge Mass: Harvard University Press, 1999).
13 Emma Garnett, "Air Pollution in the Making: Multiplicity and Difference in Interdisciplinary Data Practices," *Science, Technology & Human Values* 42(5)/2017, pp. 901–924; Emma Garnett, "Knowledge Infrastructures of Air Pollution: Tracing the In-Between Spaces of Interdisciplinary Science in Action," in: *Ethnographies and Health*, eds. Emma Garnett, Joanna Reynolds and Sarah Milton, (Cham: Palgrave Macmillan 2018), pp. 233–252.
14 Abby Kinchy, "Citizen Science and Democracy: Participatory Water Monitoring in the Marcellus Shale Fracking Boom," *Science as Culture*, 26 (1)/2016, pp. 88–110.
15 Amelia Fiske, "Dirty Hands: The Toxic Politics of Denunciation," *Social Studies of Science* 48 (3)/2018, pp. 389–413.
16 Gwen Ottinger and Elisa Sarantschin, "Exposing Infrastructure: How Activists and Experts Connect Ambient Air Monitoring and Environmental Health," *Environmental Sociology* 3(2)/2016, pp. 155–165.

as the legitimacy of intervention to improve the situation may vary.[17] Another problem is the status of sensory knowledge. In the case of smog, we are dealing with different orders of perception, in which subjective knowledge is either the basis for making a claim against public institutions or delegitimized by an expert discourse.[18] Thus, air pollution is an object of sensory politics (Spackman, Burlingame 2018) – practices of inclusion and exclusion of various orders of knowledge, highlighting certain problems in discourse or their "hermeneutical marginalisation."[19]

Our paper focuses on air quality monitoring infrastructure and controversies related to smog data collection standards and infrastructures in Poland. We are interested in two aspects: 1) relationships between different actors dealing with smog and their infrastructures in the context of metrological controversies; 2) relationships between metrological chains and ontology of the objects they refer to. In the context of this first problem, we mainly look at the dynamics and the subject of conflicts associated with measurement. In our case, the ontology of smog is related to metrological chains – different actors approach the monitoring of smog in a different way, using a variety of tools and thus allowing for its varied enactment. In order to properly capture the relationships we are interested in, we first reconstruct the map of actors involved in the controversy, then describe the conflicts around air quality monitoring, and, subsequently, point to the ontological dimension of the controversy, related to the relationship between the infrastructure, data and the objects to which they relate.

Our conclusions are based on empirical research, which consists of analysis of existing data (documents, reports, Internet press sources) and in-depth interviews with experts working in the public sector, activists, representatives of private companies and citizens (n=14). The research falls within the scope of a number of previous studies in the field of enviromental STS,[20] which deal with infrastructure and pollution data. Many of them focus on conflicts between

---

17 Gwen Ottinger, "Buckets of Resistance: Standards and the Effectiveness of Citizen Science," *Science, Technology & Human Values* 35(2)/2010, pp. 244–270.
18 Nerea Calvillo, "Political Airs: From Monitoring to Attuned Sensing Air Pollution," *Social Studies of Science* 48(3)/2018, pp. 372–388.
19 Miranda Fricher, *Epistemic Injustice. Power & the Ethics of Knowledge*, (Oxford: Oxford University Press, 2007).
20 Gwen Ottinger, "Opening black boxes. Environmental justice and injustice through the lens of science and technology studies," in: *The Routledge Handbook of Environmental Justice*, eds. Ryan Holifield, Jayajit Chakraborty, Gordon Walker, (New York: Routledge, 2017).

public institutions and citizens.[21] As we will demonstrate, many of said phenomena can also be observed in our case. In Poland, however, a completely different role than in the cases described in the literature is played by the private sector, which is not so much identified as a source of pollution, thus becoming an opponent of environmental activists,[22] but rather a competitor of public institutions in providing data attractive to citizens. As we will soon see, the great role of companies offering alternative measurement infrastructure has a bearing on the dynamics of metrological controversies and the functioning of the smog problem in public debate.

Actors and Their Infrastructures

The main public institution dealing with the state of the environment in Poland, including air quality, is the Inspectorate of Environmental Protection (IEP). The IEP consists of one central institution (Chief Inspectorate of Environmental Protection) and 17 local branches (Provincial Inspectorates of Environmental Protection). The key area of activity of the IEP is the National Environmental Monitoring, which collects data on the level of pollution in soil, water and air. The infrastructure used to measure air quality consists of a number of measurement stands and stations. The former collect data on a single selected pollutant, the latter on many different pollutants. Currently, the IEP utilizes 2187 measurement stands and about 280 stations. The stations are a key element of the infrastructure, as their outputs are used for annual air quality assessments. The measuring stations collect data using two methods: automatic and gravimetric. The first provides knowledge about the current state of the air (showing the situation from the last hour), while the second is based on the extraction of particles from the filters placed in a special sampler which sucks in air throughout the day. The data collected using the automatic method is exploited either as a support for data obtained via the gravimetric method or for informational purposes. Out of 280 stations, 190 carry out measurements in an automatic mode. Due to the growing public interest in the problem of air pollution, a few years ago the IEP

---

21 E.g. Ottinger, "Buckets of Resistance."
22 E.g. Kim Fortun, Lindsay Poirer, Alli Morgan, Brandon Costelloe-Kuehn, and Mike Fortun, "Pushback: Critical Data Designers and Pollution Politics." *Big Data & Society* 3(2)/2016, pp. 1–14; Abby Kinchy, "Citizen Science and Democracy: Participatory Water Monitoring in the Marcellus Shale Fracking Boom," *Science as Culture* 26(1)/2016, pp. 88–110; Amelia Fiske, "Dirty Hands: The Toxic Politics of Denunciation," *Social Studies of Science* 48(3)/2018, pp. 389–413.

decided to launch a web portal providing access to data collected from automatic monitoring. However, the automatic method cannot constitute the basic source of air quality data collection for the IEP. The gravimetric method is the so-called reference method. It means that only the data collected by stations equipped with gravimetric samplers can be used to perform, from the IEP point of view, the methodologically correct measurement of high accuracy and precision. The methodology of data collection, as well as its distribution, is determined by legal regulations, both at the national and European level. The IEP assesses the air quality of 11 different pollutants once a year. Poland is divided into 46 zones,[23] and for each of them a separate assessment is carried out.

Basing air quality monitoring on the reference method has important implications for the functioning of public infrastructure. First, it is an infrastructure that requires a large amount of both financial and human resources. A single monitoring station can cost as much as several thousand euros. To operate it, specialist staff and appropriate facilities are required. Sampler filters are weighed in special rooms where the right temperature is maintained. The stations themselves are checked annually by the National Reference Laboratory, a dedicated institution dealing with the quality of measurements. Secondly, because of the limited number of monitoring stations, the public infrastructure does not cover all regions of the country equally. A single zone may accommodate only several monitoring stations. The problem of "blind spots" concerns rural areas and small towns. This is why the IEP uses mathematical modelling to estimate air quality across the whole zone through point-based measurements. The modelling is performed, inter alia, on the basis of land relief data, meteorological data or data on pollution sources.

The second key actor is local self-government. Its responsibility is to shape the policy aimed at protecting air quality in the city, commune and voivodeship. Self-government authorities receive annual air quality assessments carried out by the IEP. If the number of exceedances of acceptable standards is high, the self-government is obliged to introduce the so-called Air Protection Programmes, i.e. a number of actions aimed at reducing the level of pollution. These actions may consist in limiting car traffic or introducing policies in the field of so-called thermomodernization (subsidies for the replacement of old furnaces or insulation of buildings). The inspection body of local self-government with regard to

---

23 GIOŚ (Główny Inspektorat Ochrony Środowiska), *Analiza Wybranych Epizodów*, p. 104.

environmental pollution is the municipal police, which controls, amongst other things, what is burnt in furnaces and, in the event of violations, issues fines.

Local self-government is one of the bodies to which citizens who are concerned about air quality turn up. Therefore, the approach of local authorities to monitoring and measurement infrastructure is ambivalent and largely dependent on specific circumstances. In some cases, local authorities try to provide their citizens with data on pollution when they feel that the public infrastructure is insufficient. For this purpose, local governments either turn to the IEP with a request for an additional station in a specific location or purchase sensors from private companies. The latter situation is more frequent, as without adequate resources, the IEP is often unable to set up an additional monitoring station in a specific area. By purchasing sensors from private companies, local self-government contributes to the development of alternative measurement infrastructure using different devices and methodologies from public monitoring. In other cases, however, local self-government may not be interested in the dissemination of smog knowledge. There is a tension between the interests of local self-government and local communities. In the study we came across three situations in which controversies occurred at the interface between the measurement infrastructure and the interest of a local self-government. The first concerned the use of sensors purchased from private companies to undermine the credibility of a state-owned station located in the city centre. The second was the case of health resort towns, which struggle with high concentrations of PM10 and PM2.5, which may result in the loss of the status of a health resort.[24] The third situation was related to the reluctance of some local self-governments, exposed to air pollution, to invest in measurement infrastructure indicating the scale of the problem.

The third actor are non-governmental organisations. Problems with air quality in Poland led to the emergence of the so-called Smog Alerts, a specific group of NGOs whose activities focus on the fight against air pollution. The first Smog Alert was established in 2012 in Kraków, a city facing major air pollution problems. When smog became a topic of discussion throughout the country, Alerts were established also in other towns and cities. They played an important role in raising citizens' awareness of the scale of the problem and, in several

---

24 NIK (Najwyższa Izba Kontroli), *Spełnianie wymogów określonych dla uzdrowisk* (Warszawa: NIK, 2016); WP (Watchdog Polska). "Czy w Uzdrowisku Można Odetchnąć Świeżym Powietrzem?," 2019. (17 Sep 2019) https://siecobywatelska.pl/czy-w-uzdrowisku-da-sie-odetchnac-swiezym-powietrzem/.

cases, they effectively influenced local legal regulations.[25] The Cracow Smog Alert, which has remained one of the largest organizations of its kind in Poland, is also involved in educational activities and coordinating the work of other Alert organisations. Moreover, it runs the SmogLab information portal.

Our research shows that smog-related activism can take two forms. The first is a grassroots movement involved in the fight against pollution, strongly focused on specific initiatives, but not linked to a specific political party. An example is the Cracow Smog Alert, which has thus far not been involved in a dispute over smog from political party positions. One of the characteristic features is that they are highly networked, i.e. they cooperate with a number of entities (public, commercial, civic). The second form is an initiative whose representatives are associated with a specific political party, most often in opposition to local authorities. The problem of air pollution in these forms of activism is usually one of the many issues raised against local authorities. Unlike grassroots movements, initiatives of this type tend to be autonomous – they rarely cooperate with other entities.

For anti-smog activists, infrastructure remains an important dimension of the air quality problem. Many organizations – including the largest Alert organizations in Cracow and Warsaw – cooperate with private entities offering alternative measurement infrastructure to public monitoring. Although they do not undermine the credibility and reliability of the measurements, they point out that this institution is tardy, overly official in language and detached from everyday problems of ordinary citizens. Among the accusations against public institutions are also those related to infrastructure and smog data collection. We will discuss them in detail later in the paper.

Next to IEP experts, local politicians and activists, a very important actor is the private sector. As has already been mentioned, a specific tracking culture has developed around smog, which is reflected in the growing popularity of various types of gadgets for air quality testing. Easy to install sensors checking levels of various contaminants, air purifiers or antismog masks are widely popular. The most important actor are companies that offer installation and maintenance of air quality sensors. The largest of them cooperate with Smog Alerts, which in smaller towns and cities publicize the problem of air pollution, thus encouraging local self-governments to invest in a private measurement infrastructure. Private sensors are thus becoming a cheaper and more accessible alternative to public

---

25 Mikołaj Tomaszyk, "Action Against Smog at Local Government Level in Relation to Urban Public Transport: Evidence from Selected Polish Cities," *Urban Development Issues* 55/2017, pp. 57–66.

and expensive measuring stations. The sensors used in the private infrastructure collect data using a laser particle detector, which allows automatic measurement. The detectors are manufactured on the basis of cheaper components and therefore their cost is much lower compared to IEP stations. In addition, there are no regulations in Polish law relating to the collection of data on air pollution by private entities. This means that companies enjoy a lot of freedom when it comes to the positioning of sensors and the methodology of data analysis. Companies are not obliged to comply with standards relating to the quality of data collection and analysis. Although many of them declare adherence to high standards, none of them reveal methodological and technical details, which makes the evaluation of their services significantly hindered. As a result, private infrastructure may be much denser as compared to public infrastructure, as its development is not restricted by any regulations and is simply cheaper. For instance, there are about 280 IEP stations (190 of which are automatic sensors), while Airly, the largest player on the Polish market, has distributed nearly 1800 sensors.

Aside from all of the above entities, it is possible to identify other important actors. A significant role in smog monitoring is played by international institutions, such as the World Health Organization (WHO) and the European Environment Agency (EEA). The WHO defines air quality standards for annual average concentrations of PM10, which are often quoted by the media and activists around Smog Alerts. The EEA receives annual reports from the IEP. In addition, the EEA publishes data on air pollution that has caused a stir in the Polish public discourse, which we describe below. Poland is also obliged to respect the provisions of the CAFE European Directive on air quality.

Metrological Controversies

In STS, metrology remains an important element of research.[26] The history of science knows many examples of controversy over measurement. Disputes of this kind show unstable social facts which, using the famous Bruno Latour comparison, cannot move around the social in a non-problematic way like trains do not work off their rails.[27] Examples include disputes over the methodologies for performing measurements and the admissible radiation doses after the

---

26 Alexandre Mallard, "Compare, Standardize and Settle Agreement: On some Usual Metrological Problems," *Social Studies of Science* 28 (4)/1998, pp. 571–601.
27 Bruno Latour, *Reassembling the Social: An Introduction to Actor-Network Theory* (Oxford: Oxford University Press, 2005).

Chernobyl disaster,[28] controversies over the use of the kappa coefficient to create a new classification of mental disorders,[29] and the setting of a standard for measuring adipose tissue in the human body.[30]

Smog is, in our opinion, an example of metrological controversy, which is reflected also in other studies.[31] In our case, we may distinguish three levels of controversy: 1) methodological, referring to the methods used in measuring and reliability of data; 2) analytical, related to the methods of data indexation – in the case of smog there are situations in which different actors use different indexes, presenting the same data, but interpreting it differently; 3) communicative, referring to informing about risk – in this case the concern are the thresholds for alarming about danger. The controversy surrounding air quality monitoring also makes the relations between the above mentioned actors antagonistic in character. The dynamics of the controversy is based mainly on questioning the reliability of measurements, collection, analysis and presentation of data. The participants of the controversy point to methodological errors, non-transparency of indexing methods or ideological considerations hidden behind the supposedly objective measurements. At the same time, they use the infrastructure to legitimize their position by promoting their own measurement methods and tools, convincing others about the reliability of the data they collect.

The emotions are strongest when it comes to methodological issues related primarily to data collection standards and the quality of monitoring equipment. There is a fundamental difference between public IEP stations and private infrastructure. The IEP stations operate in accordance with scrupulous methodological guidelines concerning the method of measurement or the location of sensors. Commercial entities, on the other hand, are not governed by any regulations, thus their sensors can be installed practically in any way they prefer. Representatives of the IEP publicly and in the interviews we have conducted were

---

28 Kate Brown, *Manual for Survival: A Chernobyl Guide to the Future* (New York: W.W. Norton & Company, 2019).
29 Kirk A. Stuart, and Herb Kutchins, *The Sellinf of DSM: The Rhetoric of Science in Psychiatry* (New Brunswick: Aldine Transaction, 1992).
30 Joseph O'Connell, "Metrology: The Creation of Universality by the Circulation of Particulars," *Social Studies of Science* 23(1)/1993, pp. 129–173.
31 Wenhong Chen, Fangjing Tu, and Pei Zheng, "A transnational networked public sphere of air pollution: analysis of a twitter network of PM2,5 from the risk society perspective," *Information, Communication & Society* 20 (7)/2017, pp. 1005–1023; Nerea Calvillo, "Political Airs: From Monitoring to Attuned Sensing Air Pollution," Social Studies of Science 48(3)/2018, pp. 372–388.

critical of commercially available air quality sensors. The lack of transparency on the part of private companies is problematic for officials. In their opinion, companies offering cheap sensors do not disclose details of their methods of data collection and analysis, which makes it impossible to verify its reliability. Criticism also concerns the placement of sensors. When choosing the location of a monitoring station, the IEP takes into account, inter alia, the topography and weather conditions in a given location. Location is vital with regard to data – if the sensor is located in a location with poor airflow, it will indicate higher concentrations than the sensor located in an open area and the reading will thus be misleading. The IEP representatives we spoke with encountered cases of inadequate placement of private sensors, resulting in the collection of data that misrepresented the actual level of air pollution.

An important element of the methodological dimension of the controversy is also the quality of the measurement infrastructure. The private sector tends to use cheap optical sensors. Due to the fact that many of them are not equipped with air heaters, at low temperatures they may erroneously interpret water vapour as smog and thus overestimate the results. This problem was pointed out to us both by IEP representatives and employees of one of the private companies, who thereby criticized their competition. Lack of air heaters may lead to misinterpretation of poor air quality information by data recipients. Descriptions of such occurrences have been reported in the Polish press. For example, in Kalwaria Zebrzydowska, a small town in southern Poland, the particulate matter standards were exceeded on one winter day by several thousand percent. Residents shared data from private sensors in social media, and the incident was also publicized by the Kalwaria Smog Alert. The spokesman of the City Hall, after the intervention of the IEP, in an official statement had to convince the residents that nothing extraordinary is happening and that the fault lies with the defective sensors. As a representative of one of the companies told us, such situations happen quite often: "when residents see a red indicator, they call the mayor."

Aside from the quality and methodology of measurement, the methods of presenting data in the form of air quality indices are also the subject of controversy. Indices are used to translate concentration readings into concrete interpretations. The problem with monitoring smog in Poland is that there are different indices that adopt different concentration ranges. The IEP uses its own index, the Polish Air Quality Index (PAQI). Meanwhile, many European countries use the Common Air Quality Index (CAQI), which covers substantially different ranges. Some private companies in Poland use the modified version of CAQI, and often define their own concentration levels. For instance, the CAQI index has five levels: "very low," "low," "medium," "high," and "very high." On the other hand, the Airly CAQI index, created by the largest Polish company offering air quality sensors, has as many as seven levels: "very low," "low," "medium," "high," "very high," "extreme" and "airmageddon!." These differences may, of course, give rise to interpretation problems, since potentially the same reading may be interpreted differently depending on the index used.

The difference in the indices used is raised as an argument against public institutions. This issue is voiced by the Smog Alerts, arguing that the index used by the IEP may mislead the public. The PAQI index used by the IEP assumes six degrees: "very good," "good," "moderate," "satisfactory," "bad" and "very bad." Controversy is sparked by the "satisfactory" level, as in the case of PM10 it indicates ranges from 101.1 µg/m3 to 141 µg/m3. The word "satisfactory" suggests that the situation is not so bad, whereas in the opinion of activists such concentrations should be considered as indicating poor air quality.[32] Criticising the IEP, the Alerts compare PAQI to CAQI – within the latter a level of 140 µg/m indicates air with a "high" concentration of harmful substances.

The final dimension of the controversy is of a communicative nature. It relates to the question of communicating the problem of smog. The issue is related to air quality standards and alert and risk notification thresholds. Air quality standards are defined by European Parliament directives and find reflection in Polish legislation. These standards do not specify the level above which smog is a health problem. However, the directives set admissible levels. For PM10, this is 50 µg/m3 on a 24-hour scale and 40 on an annual basis, while PM2.5, which is more harmful to health, has a standard set only as per year – 25 µg/m3. We can compare it to the standards suggested by the WHO. The WHO recommendations specify the PM2.5 limit also for daily averages. It is worth stressing here that European standards are binding on all Member States, while the WHO standards are only suggestions based on the knowledge about the harmfulness of smog that has been accumulated to date. Neither the European Union nor the WHO defines standards for hourly measurements.

These issues become relevant when information on the current state of air in various Polish cities is presented in public discourse. This is particularly true during winter, when the problem of smog grows and more often resounds in the media. Air quality standards are invoked to highlight the importance of the problem and to alert the public opinion. On the one hand, the media use automatic monitoring data, which provides information about the state of

---

32 KAS (Krakowski Alarm Smogowy), *Co Wiemy o Smogu? Informowanie o zanieczyszczeniu Powietrza w Polsce* (Kraków: Krakowski Alarm Smogowy, 2015), p. 40.

pollution from the last hour, but at the same time they use standards that refer to average concentrations during the 24-hour period. Another practice is to refer to WHO standards, which are more stringent and, in the case of PM2.5, define a daily standard. With this approach, data can be presented in a more alarming way.

From the perspective of the IEP, such practices are obviously unauthorized, as they are not based on legally sanctioned ways of interpreting data and may, similarly to low quality measurement equipment or inappropriate location of the sensor, mislead the public. Because of discrepancies in standards of data interpretation, this institution sometimes becomes a subject of criticism from the media and the public opinion. An example is the situation in 2017, when the EEA published data showing that Poles breathe with the worst air in the European Union. At exactly the same time, data from IEP stations indicated that the situation was not so bad – while the European map covered Poland with a blood red colour, indicating very poor air quality, the IEP map was orange-green. This difference became the subject of controversy. The press blamed the IEP for showing a distorted picture of the problem. In reality, the databases contained different data. The EEA map featured data with daily concentrations, while the IEP map the concentrations of the last hour. From the methodological point of view, these data are not comparable. Nevertheless, at the level of public discourse, these nuances remained unnoticed, provoking criticism of public institutions.

The problem of communicating the smog threat is also related to alert and notification levels. When a certain average daily concentration level is exceeded, the IEP is obliged to inform the public through the local Crisis Management Centre and local authorities of the concentration that may be a health hazard or, in the event of higher concentrations, to warn that even a short-term exposure to contaminated air may adversely affect human health. The problem is that the EU does not define a concentration range for PM10 and PM2.5 that would require notification or alerting, leaving it to the discretion of each Member State. As a result, there are significant differences between them.

These differences can be seen, for example, when we compare Poland with France. According to French criteria, the alert level of a high daily concentration of PM10 is 80 μg/m3, while in Poland it is as high as 300 μg/m3. As compared to other countries, it appears that Poland uses some of the highest alarm levels. As a result, activists raise the accusation of the problem being trivialized. The Smog Alerts point out that if the French criteria were adopted, the alarm would remain in place in many Polish cities for several weeks. The problem of an excessively liberal approach to the level of air pollution was addressed by the Polish Smog Alert and Greenpeace, on whose initiative a petition signed by 10 thousand people was sent to the Ministry of the Environment asking to reduce the alarm threshold to the level of 100 µg/m3. The Ministry refused to agree to this. The Ministry's letter sent to Greenpeace reads:

> The introduction of an alarm level ... below 300 µg/m3 or even 100 µg/m3 would be ineffective. It would result in the necessity of frequent announcing the alarm status. Due to the scale of the phenomenon, it may cause society to quickly become accustomed to this state of affairs and not react to a situation where concentrations of PM10 particulate matter are repeatedly exceeded and fail to treat the exceedance of air quality standards as an emergency.

It is clear that smog is a controversial issue not so much in terms of its impact on health – in Poland nobody seems to argue about it – but in terms of infrastructure and standards. At this point, one may ask what particular entities seek to achieve by engaging in the controversy? The IEP, as a public and largely expert institution, perceives its role as a guardian of the quality and reliability of pollution data. However, from a wider perspective on the whole public sector, air quality monitoring is in many cases used as a lever for managing the public discourse. Given the fact that smog has become a hot topic of social debate, strict methodological requirements, restrictive measurement standards or expensive infrastructure often contribute to the trivialization of the problem and suppression of emotions. This can be seen very clearly in the case of alert and notification thresholds. The reply of the Ministry of the Environment is an excellent example of the fact that the smog concentration level is established with consideration of social emotions, and not with respect to the substantive issues. Thus, on the one hand, the metrological chain used by public institutions creates what Andrew Barry[33] describes as the metrological zone, an area where, by establishing a single measurement standard, it is possible to objectify a phenomenon and establish a reference point for other methods of measurement or analysis. In the case of smog, the public metrological zone encompasses expensive, hard-to-access resources that can only be utilized by defined actors. The threshold for entering the area where reliable and credible knowledge is produced is set very high. The IEP reminds the public and the companies that provide cheap sensors that it is the only institution that guarantees reliable and methodologically correct knowledge on the subject of smog. On the other hand, however, standards and infrastructure are to produce data and interpretations that not so much serve the purpose of understanding reality as to reinforce a politically safe image of the

---

33 Andrew Barry, "Technological zones," *European Journal of Social Theory* 9(2)/2006, pp. 239–253.

world. The world in which smog is a problem, but not serious enough to become an object of constant social panic. Ingmar Lippert[34] calls this the production of comfort. It seems that this function is performed by high alert and notification thresholds which are set in such a way as not to overly electrify public opinion.

From the private sector's perspective, the controversy around air quality monitoring constitutes an opportunity to establish oneself as an entity capable of creating reliable and, above all, easily accessible knowledge on air pollution. We have indicated before that the private sector often becomes a partner for local self-government and activists, offering support that is sometimes impossible to obtain from the IEP. Below we will discuss also the ontological issue, because due to the different enactment of smog in the metrological chain, the private sector more effectively meets the social demand for air quality data as compared to the public sector. In this context, let us highlight the conflicting relationship between the public and private sectors. Interestingly, in our interviews, representatives of companies fitting cheap sensors did not speak negatively about public institutions, pointing out that their infrastructure is not competitive to the IEP, but rather provides a necessary supplement to it. This cannot be said of the IEP, for which the private sector is a problem, as it offers poor quality smog data. The activities of the IEP in this context can be classified as policy-bridging practices,[35] i.e. using standards and infrastructures in such a manner as to discredit the standards and infrastructure of other entities, setting boundaries between legally valid and invalid knowledge. In the literature describing the clash of different perspectives on environmental pollution, public institutions have often been described as intermediaries in conflicts between the industry and citizens,[36] or as parties delegitimising knowledge produced by activists.[37] In this case, it is different – it is the private sector, not just citizens and activists, that is positioned by the IEP as an actor producing unreliable and sometimes harmful knowledge about smog.

---

34 Ingmar Lippert, "Failing the Market, Failing Deliberative Democracy: How Scaling Up Corporate Carbon Reporting Proliferates Information Asymmetries." Big Data & Society 3 (2)/2016, pp. 1–13.
35 Ottinger, "Buckets of Resistance."
36 E.g. Fortun, Poirer, Morgan, Costelloe-Kuehn, and Fortun, "Pushback."
37 Ottinger, "Buckets of Resistance."

Smog Ontologies

The relationship between standards and infrastructures and object ontologies is a topic frequently discussed in STS. We can mention here the works of Susan Leigh Star and James R. Griesemer or Annemarie Mol.[38] The above described metrological controversy around smog can also be described from an ontological perspective. In this context, different metrological chains would not only produce different knowledge of smog, but would also apply distinct definitions of the very phenomenon they concern.

The key difference pertains to temporality, as it is related to the time to which the pollution data refer.[39] The two types of measurement discussed above – automatic and gravimetric – yield, respectively, information on the last hour's concentration level or the daily average. As a reminder, automatic measurement is used mainly by the private sector, as well as the IEP as a support for other methods, while gravimetric measurement is used by the IEP. At the same time, these are two different metrological chains– the first consists of cheap, easy to use and operate sensors, the second comprises costly and requiring expensive infrastructure measurement stations. The first chain consists of a rich network of sensors, showing the level of concentrations in specific locations. The network of public measurement stations is less developed, which means that the IEP has to use mathematical modelling to assess the condition of the air in a larger area on the basis of point measurements.

It is interesting to note that the varying temporality of the data entails a difference in terms of what smog is and what can be done about it. In the case of automatic monitoring, data recipients can learn about the current state of air pollution, making smog an object situated in a specific place and time. Thanks to the data showing the situation from the last hour the threat associated with polluted air changes into something concrete and experienceable for the individual body. We should not be surprised that these data, as our respondents pointed out, are used to make decisions about everyday activities, such as taking a walk with a dog. The situation is different with monitoring based on the gravimetric method. Data showing average daily concentrations from the point of view of

---

38 Susan L., Star, and James R. Griesemer. "Institutional Ecology, 'Translations' and Boundary Objects: Amateurs and Professionals in Berkeley's Museum of Vertebrate Zoology, 1907-39," *Social Studies of Science* 19(3)/1989, pp. 387–420; Annemarie Mol, *Body Multiple: Ontology in Medical Practice* (Durham Duke University Press London, 2002).
39 Ottinger and Sarantschin, "Exposing Infrastructure."

ordinary citizens are less effective than those from the last hour. However, they are necessary for periodic air quality assessments, and this is the main objective of the IEP. Gravimetric data are mainly used to make administrative decisions concerning larger groups of people and larger spaces. In this case, smog is something different than in the case of automatic monitoring because it is presented in a different time and spatial scale. Smog becomes a biopolitical threat.[40] affecting the population living in a certain territory over a longer period of time. The knowledge about it is statistical in nature and the actions taken in respect to it take the form of risk management.

The ontological difference can also be seen in relation to sensory experience. Different metrological chains support different regimes of perception[41] and thus legitimize different orders of knowledge. With a distinction between laboratory and sensory perception,[42] we could say that automatic monitoring supports the latter, while monitoring using the gravimetric method supports the former. Identification of environmental threats through visual recording of the damage that a given substance or poor air quality may cause to an organism is a common practice of citizens who experience negative effects of environmental problems.[43] Smog is a good example here as it can be felt and seen. As one committed citizen told us, "if you could see these fumes hovering over the city, if you could see how they smell, you can see with the naked eye, it is thick air." An extensive network of automatic sensors not only situates the problem of smog in a specific time and space context, but also provides a scaffold for the physical experience of smog, as it links sensory experience to data. When air quality appears to be poor, for example when there is a characteristic mist or unpleasant smell, a person equipped with a cheap sensor or an Airly service user can verify whether their impression is correct.

In expert ontology, subjective, qualitative and situated knowledge about smog is seen as part of scientific and quantitative knowledge. Smog is not something

---

40 Michel Foucault, "The birth of biopolitics," in: Michel Foucault, *Ethics: Subjectivity and Truth*, ed. Paul Rabinow, (The New Press New York, 1997), pp. 73–80.
41 Max Liboiron, Manuel Tironi, and Nerea Calvillo, "Toxic Politics. Acting in a Permanently Polluted World," *Social Studies of Science* 48(3)/2018, pp. 331–349.
42 Christy Spackman, Barry A. Burlingame, "Sensory Politics: The Tug-of-War Between Portability and Palatability in Municipal Water Production," *Social Studies of Science* 48(3)/2018, pp. 350–371.
43 Jennifer Gabrys, "Citizen Sensing, Air Pollution and Fracking: From 'Caring About Air' to Speculative Practices of Evidence Harm," *The Sociological Review Monographs* 65(2)/2017, pp. 172–192.

directly observable, but it is contained in tables, charts or chemical diagrams. And such an enactment of smog is possible thanks to laboratory infrastructure. Let us take an example. A resident of Grudziądz, one of the most polluted cities in northern Poland, complained to us that the municipal police did not take appropriate action against his neighbour due to the fact that suspiciously looking smoke was rising from his chimney. On the other hand, another respondent, a city guard, clearly stated that "neither the colour of the smoke nor the smell is evidence in the case." In order to issue a fine an analysis of the chemical composition of the hearth (which is carried out in a special laboratory) is required, as only burning fuel that is illegal from the perspective of the law can meet with an appropriate reaction and sanction. Similar problems were discussed with us by the representatives of the IEP, who are approached by citizens with requests for intervention. These requests are often based on sensory or visual experiences. For the IEP this cannot, however, be decisive. The main source of certain knowledge is the rigorous methodology defined by EU and national regulations, not the testimonies of citizens: "Our data needs to be verified. These are data for evaluation, data that go to the European Commission. These are figures that are later used in air quality management. And we cannot allow these data to raise any doubt." Such verified data come only from monitoring based on a gravimetric method, the quality of which is guaranteed by appropriate staff and technological facilities.

It is worth stressing that the difference between the above mentioned expert and civic ontology does not coincide with the division into quantitative and qualitative. According to Ottinger and Sarantschin,[44] experts, as well as citizens politically committed to improving air quality, to some extent move within a similar framework, as they rely on quantitative monitoring. This means that a purely qualitative (e.g. sensory) approach to smog is perhaps impossible without a quantitative prosthesis, and the differences between perspectives can be seen not so much in the use of quantitative data as in how it is done. Ottinger and Sarantschin indicate that one of the key differences in the use of the data concerns their temporality. Experts, more than activists, rely on mathematical models, are primarily focused on risk assessment and are interested in the long-term rather than short-term effects of air pollution on the body. On the other hand, Fortun et al[45] believes that a bottom-up, civic view of environmental problems using pollution data is more hermeneutical (because it takes into account not only what

---

44 Ottinger and Sarantschin, "Exposing Infrastructure."
45 Fortun, Poirer, Morgan, Costelloe-Kuehn, and Fortun, "Pushback."

certain data mean, but also for whom and in what circumstances they become relevant) and situated (because they relate to a specific context). In Ottinger's article[46] on the bucket brigades movement, the difference between measuring pollution in the short and long term was a source of a conflict between activists and experts. In this case, activists accused public institutions carrying out environmental monitoring that focusing only on the problem of a long exposure to negative chemical agents underestimates the problem. Therefore, they not only legitimized their claims, but deliberately referred the results of continuous measurements to the standards for average measurements. This practice did not result from their methodological ignorance, but was intended to draw experts' attention to the flawed standards and thus legitimize another ontological policy in which polluted air may directly and in a short period of time adversely affect health.

Otinger's studies are particularly important because they show that multiple ontologies of objects can frame relations between different entities not in the form of boundary-bridging – as is the case with boundary objects[47] – but in the form of a conflict. From this perspective, conflicts of activists and citizens with public institutions and experts are not only a manifestation of metrological controversy, but also a clash of diverging ontologies. Also in the case of smog, there is no coordination, as its ontological distinctness does not allow to agree on perspectives. In our case, the following quotation testifies to this. It comes from an interview with a city councilor, who is also a pro-ecological activist:

> So they have a guideline that it's not supposed to be closer than some meters from the chimney, maybe a hundred, I don't remember now, so these sensors are probably positioned in such a way that they show the average correct level, right? But what if my child leaves home in Bydgoski district, goes to a kindergarten in Bydgoski district, another goes to school in Bydgoski district, we go to a shop in Bydgoski district and generally move around this district. I'm breathing that air, not averaged air, right? In this sense I'm not happy with it, because it tells me what kind of air councilmen breathe, when they're somewhere near the City Hall, right? And not everyone breathes that kind of air.

For this person, smog is something concrete and situated, and the knowledge about it should be adapted to the conditions of everyday life. From this

---

46 Ottinger, "Buckets of Resistance."
47 Susan L. Star, "This is Not a Boundary Object. Reflections on the Origin of a Concept," *Science, Technology & Human Values* 35(5)/2010, pp 601–617; Star, "Institutional Ecology."

perspective, the placement of sensors and adherence to high standards (i.e. elements of expert ontology and public metrological chain) does not meet his expectations.

On the other hand, however, from the perspective of an expert institution such as the IEP, it is impossible to meet such social expectations. There are not enough funds, because expanding the automatic monitoring network ("placing a sensor on every street") would generate too high costs. There is also the question of understanding what pollution is. As one of the officials told us, "people look at moments," i.e. at the readings from the automatic measurement, without understanding the time necessary to average the measurements. Due to the variability of atmospheric conditions (a high reading at one hour may be radically different at another hour due to wind), as well as the lack of regulations concerning the short-term impact of smog on the organism, such an individualized and focused on a specific place and time perspective is criticized by the IEP as unfounded.

The plurality of smog ontologies can also provide a perspective on the role of the private sector. Using a cheap and easily accessible measurement infrastructure based on automated monitoring, companies not only offer a product which is more attractive compared to what is offered by public institutions, but also enact smog within the same ontological order as citizens. This is best seen when we look at how Airly, Poland's largest private measurement infrastructure company, describes its data. A map presenting some of their sensors shows not only a specific indication of the concentration of harmful substances, but also short descriptions that provide social positioning of the data, e.g: "Wonderful air! Breathe deeply," "Wonderful air for a walk in the park!," "Poor air quality! You had better postpone the walk to another day," "There's smog hovering over this location. We recommend that you limit your outdoor activities." From this perspective, the methodological defects of private infrastructure, which the IEP points out, are not very important because, as Pantzar and Ruckenstein write, "once the measured data are contextualized in the everyday, accuracy might no longer be their most important aspect or value."[48]

Moreover, Airly presents smog data using specific visualization methods. On the company's website one can find a map showing concentration levels from specific locations. The aforementioned seven-step index uses a colour palette: from green indicating low smog level to dark red indicating extremely poor air quality. A characteristic index-coloured mist spreads around the coloured

---

48 Mika Pantzar and Minna Ruckenstein, "Living the Metrics: Self-tracking and Situated Objectivity." *Digital Health* 3/2017, pp. 1–10.

dots. It looks as if the object of visualization is not only the point reading, but also the air around it. From the methodological point of view, this form of data presentation is not legitimate – automatic monitoring only collects data at the point of the sensor's location. However, the method of visualization chosen by Airly belongs to the order of perception, where smog is something situated and sensually experienced, such as air, smoke or fog. As a result, it is closer to the way in which smog is enacted in everyday life practices.

Conclusions

In our research we focused on the relationship between the actors involved in the controversy over air quality monitoring and the relationship between the smog ontology, infrastructures and standards. We have identified different regimes for developing knowledge on air pollution, which are based on different ways of collecting, analysing and presenting data and thus enactment of objects that may be subject to different practices. In the context of ontology, we pointed to two important factors – temporality and approach to sensory knowledge – which also differentiate the perspective of actors involved in the controversy.

The characteristic feature of our analysis is that it primarily focuses on the relationship between public institutions responsible for environmental protection and the private sector, which, at the expense of accuracy and scientific reliability, offers users an alternative, more accessible and more suited to everyday needs measurement infrastructure. The weakness of this approach is that in our analysis we devoted less attention to the role of activism for the improvement of air quality, which is extremely developed in Poland by the aforementioned Smog Alerts. This certainly requires further research. At the same time, this perspective has highlighted the role of companies in the controversy about smog. It is important for several reasons. Firstly, companies producing cheap sensors are becoming an important player in the field of air quality monitoring and are therefore a natural competitor for public institutions. The effect of the rapid development of the private sector is the growing role of alternative measurement infrastructure, which may in the future become the main reference point for users (the IEP also has automatic sensors in its network, but not as many as private companies). This creates both an opportunity for greater dissemination of information on air quality as well as a risk of producing unreliable knowledge. Moreover, it may also create the need to integrate different systems and infrastructures to bridge the gap between different ontologies. Secondly, data on air quality is subject to capitalization in the Polish context. This does not only concern the drawing of profits from the fact that Polish society has recently taken such a massive interest in air quality. Private infrastructure is connected with public infrastructure by dependency relations. Cheap sensors are calibrated at IEP stations, and data from the latter, as data belonging to the public domain, are rendered available in private applications. This leads to an interesting interception by private capital. The controversy over pollution in certain circumstances does not have to happen only between institutions and citizens, but may also have its political and economic background and be related to the neoliberal takeover of public resources. This interesting topic has emerged in our research and also deserves to be further developed. Thirdly and finally, the question may be posed as to whether the increasing role of the private sector is replacing practices in the field of citizen science or citizen sensing. While in Poland there exist initiatives of this kind in relation to smog, in our talks, sensors supplied by private companies were indicated as tools for collecting data outside the public sphere, rather than devices constructed on one's own. In the analyses of citizen sensing practices, an expression of disagreement with the way public institutions operate consists in civic activation, which is expressed by constructing sensors or independent data collection.[49] In our case, however, we are dealing with a civic void that is filled by private entities. Of course, this is not always the case, as exemplified by the Smog Alerts operating in large Polish cities. However, we must take into account the possibility of a different trajectory – the social interest in data may not lead to citizen sensing practices, but may become a subject of capitalization by the private sector.

Bibliography

Barry, Andrew. "Technological zones." *European Journal of Social Theory* 9(2)/2006, pp. 239–253.

Bowker, Geoffrey, and Susan L. Star. *Sorting Things Out: Classifications and Its Consequences*. Cambridge: MIT Press, 1999.

Brown, Kate. *Manual for Survival: A Chernobyl Guide to the Future*. New York: W.W. Norton & Company, 2019.

Calvillo, Nerea. "Political Airs: From Monitoring to Attuned Sensing Air Pollution." *Social Studies of Science* 48(3)/2018, pp. 372–388.

Chen, Wenhong, Fangjing Tu, and Pei Zheng. "A transnational networked public sphere of air pollution: analysis of a twitter network of PM2,5 from the risk

---

49 Gabrys, "Citizen Sensing."

society perspective." *Information, Communication & Society* 20(7)/2017, pp. 1005–1023.

EEA (European Environmental Agency). *Air Quality in Europe – 2018 Report*. Luxembourg: Publications Office of the European Union, 2018.

Fiske, Amelia. "Dirty Hands: The Toxic Politics of Denunciation." *Social Studies of Science* 48(3)/2018, pp. 389–413.

Fortun, Kim, Lindsay Poirer, Alli Morgan, Brandon Costelloe-Kuehn, and Mike Fortun. "Pushback: Critical Data Designers and Pollution Politics." *Big Data & Society* 3(2)/2016, pp. 1–14.

Foucault, Michel.. "The birth of biopolitics." In: Michel Foucault, *Ethics: Subjectivity and Truth*, edited by Paul Rabinow, New York: The New Press: 1997, pp. 73–80.

Fricher, Miranda. *Epistemic Injustice. Power & the Ethics of Knowledge*. Oxford: Oxford University Press, 2007.

Gabrys, Jennifer, Helen Pritchard and Benjamin Barratt. "Just Good Enough Data: Figuring Data Citizenships Through Air Pollution Sensing and Data Stories." *Big Data & Society* 3(2)/2016, pp. 1–14.

Gabrys, Jennifer. "Citizen Sensing, Air Pollution and Fracking: From 'Caring About Air' to Speculative Practices of Evidence Harm." *The Sociological Review Monographs* 65(2)/2017, pp. 172–192.

Garnett, Emma. 2016. "Developing a Feeling for Error: Practices of Monitoring and Modelling Air Pollution Data." *Big Data & Society* 3 (2): 1–12.

Garnett, Emma. "Air Pollution in the Making: Multiplicity and Difference in Interdisciplinary Data Practices." *Science, Technology & Human Values* 42(5)/2017, pp. 901–924.

Garnett, Emma. "Knowledge Infrastructures of Air Pollution: Tracing the In-Between Spaces of Interdisciplinary Science in Action." In: *Ethnographies and Health*, edited by Emma Garnett, Joanna Reynolds, and Sarah Milton. Cham: Palgrave Macmillan: 2018, pp. 233–252.

GIOŚ (Główny Inspektorat Ochrony Środowiska). *Pyły Drobne w Atmosferze. Kompendium wiedzy o zanieczyszczeniu powietrza pyłem zawieszonym w Polsce*. Warszawa: Główny Inspektorat Ochrony Środowiska. 2016.

GIOŚ (Główny Inspektorat Ochrony Środowiska). *Analiza Wybranych Epizodów Wysokich Stężeń pyłu PM10 z lat 2013–2016*. Warszawa: Główny Inspektorat Ochrony Środowiska, 2017.

KAS (Krakowski Alarm Smogowy). *Co Wiemy o Smogu? Informowanie o zanieczyszczeniu Powietrza w Polsce*. Kraków: Krakowski Alarm Smogowy, 2015.

Kinchy, Abby. "Citizen Science and Democracy: Participatory Water Monitoring in the Marcellus Shale Fracking Boom." *Science as Culture* 26 (1)/2016, pp. 88–110.

Kirk, Stuart A. and Herb Kutchins. *The Sellinf of DSM: The Rhetoric of Science in Psychiatry*. New Brunswick: Aldine Transaction, 1992.

Knorr-Cetina, Karin. *Epistemic Cultures. How the Sciences Make Knowledge*. Cambridge Mass.: Harvard University Press, 1999.

Latour, Bruno. *Science in Action. How to Follow Scientists and Engineers Through Society*. Cambridge Massachusetts: Harvard University Press, 1987.

Latour, Bruno. *Reassembling the Social: An Introduction to Actor-Network Theory*. Oxford: Oxford University Press, 2005.

Liboiron, Max, Manuel Tironi, and Nerea Calvillo. "Toxic Politics. Acting in a Permanently Polluted World." *Social Studies of Science* 48(3)/2018, pp. 331–349.

Lippert, Ingmar. "Failing the Market, Failing Deliberative Democracy: How Scaling Up Corporate Carbon Reporting Proliferates Information Asymmetries." *Big Data & Society* 3(2)/2016, pp. 1–13.

Lupton, Deborah. *The Quantifed Self*. Cambridge: Polity Press, 2016.

Mallard, Alexandre. "Compare, Standardize and Settle Agreement: On some Usual Metrological Problems." *Social Studies of Science* 28(4)/1998, pp. 571–601.

Mol, Annemarie. "Ontological Politics. A Word on Some Questions." *The Sociological Review* 47(1)/1999, pp. 74–89.

Mol, Annemarie. *Body Multiple: Ontology in Medical Practice*. Durham-London: Duke University Press, 2002.

NIK (Najwyższa Izba Kontroli). *Spełnianie wymogów określonych dla uzdrowisk*. Warszawa: NIK, 2016.

O'Connell, Joseph. "Metrology: The Creation of Universality by the Circulation of Particulars." *Social Studies of Science* 23, 1/1993, pp. 129–173.

Ottinger, Gwen, and Elisa Sarantschin. "Exposing Infrastructure: How Activists and Experts Connect Ambient Air Monitoring and Environmental Health." *Environmental Sociology* 3(2)/2016, pp. 155–165.

Ottinger, Gwen. "Buckets of Resistance: Standards and the Effectiveness of Citizen Science." *Science, Technology & Human Values* 35(2)/2010, pp. 244–270.

Ottinger, Gwen. "Opening black boxes. Environmental justice and injustice through the lens of science and technology studies." In *The Routledge Handbook of Environmental Justice*, edited by Ryan Holifield, Jayajit Chakraborty, Gordon Walker. New York: Routledge, 2017.

Pantzar, Mika, and Minna Ruckenstein. "Living the Metrics: Self-tracking and Situated Objectivity." *Digital Health* 3/2017, pp. 1–10.

PAS (Polski Alarm Smogowy). *Smog Alert? Have not Heard of It… Polish Iron Lungs*. Kraków: Polski Alarm Smogowy 2015.

Spackman, Christy, Barry A. Burlingame. "Sensory Politics: The Tug-of-War Between Portability and Palatability in Municipal Water Production." *Social Studies of Science* 48(3)/2018, pp. 350–371.

Star, Susan L. "This is Not a Boundary Object. Reflections on the Origin of a Concept." *Science, Technology & Human Values* 35(5)/2010, pp. 601–617.

Star, Susan L. and James R. Griesemer. "Institutional Ecology, 'Translations' and Boundary Objects: Amateurs and Professionals in Berkeley's Museum of Vertebrate Zoology, 1907–39." *Social Studies of Science* 19(3)/1989, pp. 387–420.

Tomaszyk, Mikołaj. "Action Against Smog at Local Government Level in Relation to Urban Public Transport: Evidence from Selected Polish Cities." *Urban Development* Issues 55/2017, pp. 57–66.

WP (Watchdog Polska). "*Czy w Uzdrowisku Można Odetchnąć Świeżym Powietrzem?*," September 17, 2019, https://siecobywatelska.pl/czy-w-uzdrowisku-da-sie-odetchnac-swiezym-powietrzem/ (2 April 2022).

Maria Lompe

Learning to Breath: Issue Mapping on the Smog Controversy in Poland Using the Web 2.0

Abstract: This article explores the controversy about smog in Poland based on the media messages that were most shared on social media in late 2019 and early 2020. Issue mapping, which is rooted in Science and Technology Studies, was used to locate the most widely discussed threads of the smog debate in Poland. Based on the data collected during this period, it was possible to reconstruct the main actors who participated in the debate during this time and to locate the most frequently addressed issues related to air pollution in Poland. The study also allowed to reconstruct the dynamics of the debate and to reflect on the way the content was disseminated in social media.

Keywords: smog, controversy, issue mapping, Web 2.0, science technology studies

The study of controversies is one of the primary areas of interest for researchers associated with the Science Technology Studies field.[1] It stems from the special attention paid to such moments of scientific practice when uncertainty dominates. This is when we can observe the process of knowledge creation, the negotiation between diverse actors, and often witness how knowledge is stabilized.[2] However, sometimes, despite the development of scientific consensus, basic assumptions about an issue are still challenged in public debate. This is the case with global warming denial,[3]

---

1 Chandra Mukerji, "Controversy studies," in: *The Blackwell Encyclopedia of Sociology*, ed. Georg Ritzer (Oxford: John Wiley & Sons, 2007), pp. 784–788; Ewa Bińczyk, "Doktor Golem? Badania kontrowersji w nurcie studiów nad nauką i technologią wobec statusu nauk medycznych, wybranych jednostek chorobowych i zmieniającej się roli eksperta," *Przegląd Filozoficzno-Literacki*, Vol. 1, No. 46 (2017), pp. 285–306.
2 Harry Collins, Trevor Pinch, *The Golem: What Everyone Should Know about Science*, (Cambridge: Cambridge University Press, 1993); Thomas Kuhn, *The Structure of Scientific Revolution*, (Chicago: University of Chicago Press, 1962).
3 Tomasso Venturini, Daniele Guido, "Once Upon a Text: an ANT Tale in Text Analysis," *Sociologica* Vol. 3, (2012), pp. 1–16; Richard Rogers, Noortje Marres, "Landscaping climate change: a mapping technique for understanding science and technology debates on the World Wide Web," *Public Understanding of Science*, Vol. 9, (2000), pp. 1–23; Sabine Niederer, "Global warming is not a crisis!: Studying climate change skepticism on the Web," *European Journal of Media Studies*, Vol. 2, No.1 (2013), pp. 83–112.

controversies about food technology,[4] biofuels,[5] tobacco smoking,[6] or public health[7]

In response to such discussions, over the past two decades new methods of controversy research have been developed that use digital tools. They no longer focus only on investigating scientific controversies, but much more often on investigating publicly debated issues.[8] This approach does not treat digital activity merely as a separate space of "virtual society,"[9] but rather as markers of current trends, modes of communication, and group formation processes that grow out of digital environments. The study of Controversies using digital methods relies primarily on World Wide Web data, using various types of scrappers to extract data from social networks, websites, blogs, forums, and different types of databases.

An example of such a controversy is the smog debate in Poland. Despite the development of a scientific consensus,[10] and actions taken at both international

---

4   Gerald Beck, Cordula Kropp, "Infrastructures of risk: A mapping approach toward controversies on risks," *Journal of Risk Research*, Vol. 14, No. 1 (2011), pp. 1–16.
5   Astid Mager, Jenny Eklöf, "Technoscientific Promotion and Biofuel Policy: How the Press and Search Engines Stage the Biofuel Controversy," *Media Culture & Society*, Vol. 35, No. 4 (2013), pp. 454–471.
6   Naomi Oreskes, Erik M. Conway, *Merchants of Doubts. How a Handful of Scientist Obscured the Truth on Issues from Tabaco Smoke to Global Warming*, (New York: Bloomsbury Press, 2010).
7   Federico Germani, Nikola Biller-Andorno, "The anti-vaccination infodemic on social media: A behavioral analysis," *PLoS ONE*, Vol. 16, No. 3 (2021).
8   Tomasso Venturini, "Diving in Magma: How to Explore Controversies with Actor – Network Theory," *Public Understanding of Science*, Vol 19, (2010), pp. 258–273. Tomasso Venturini, "Building on faults: how to represent controversies with digital methods," *Public Understanding of Science*, Vol. 21, (2012), pp. 796–812; Noortje Marres, "Why map issues? On Controversy Analysis as a Digital Method," *Science, Technology, & Human Values*, Vol. 40, No. 5 (2015), pp. 655–686; Marres N., Moats D, "Mapping Controversies with Social Media: The Case for Symmetry," *Social Media + Society*, Vol. 1, (2015), pp. 1–17.
9   Richard Rogers, *Doing Digital Methods*, (Los Angeles: Sage, 2019), p. 14.
10 PAS, "Poziomy informowania i alarmowe," <https://www.polskialarmsmogowy.pl/polski-alarm-smogowy/smog/szczegoly,poziomy-informowa-nia-i-alarmowe,19.html>, (07.02.202). WHO, "Ambient air pollution," <https://www.who.int/airpollution/data/cities/en/>, (05.01.2020). WHO, "Household air pollution and health," <https://www.who.int/news-room/fact-sheets/detail/household-air-pollution-and-health>, (01.05.2020).

and local levels, still every year during the winter we can follow debates on how to deal with smog in Poland. Since 2011, discussions about smog in Poland have gained intensity,[11] which also resulted in the appearance of the first controversies on this topic in the public space. The problem of air pollution in Poland is associated with phenomena as distant as a coal-burning economy, the questioning of anti-smog laws, or political camps identifying themselves with the idea of fighting for clean air in cities. The methodology of air pollution measurements is often questioned,[12] resulting in the creation of networks of amateur dust meters or the denial of the existence of the problem. Coal companies accuse the government of collaborating with the gas lobby, while clean air activist movements accuse the government of collaborating with the coal lobby. Smog has been a recurring theme in Polish media discourse for several years, but the harmfulness of air pollution continues to be discussed, and proposed solutions to the smog problem are widely debated.

Methodology

This article is an attempt to locate the most widely discussed smog issues in the Polish media debate, from the perspective of Science & Technology Studies (STS), using digital methods. In the chosen method, the Google browser was treated "as an epistemological machine,"[13] which in research using digital tools is treated as a "provider" of major social trends. The nature of the controversiality of a piece of information, assumed in the study, was determined by the reaction it elicited.[14] The aim of the study was not to map the widest possible area of news addressing the topic of smog in Poland, but rather those news that were widely commented on social media. The intensity of a given controversy was determined by the number of shares and reactions for a given article on Facebook. Articles that were published in mainstream media but did not gain any reactions on Facebook were therefore not considered for this study. In

---

11 Smoglab, "Smog był zawsze, czemu mówimy o nim teraz?," <https://smoglab.pl/smog-byl-zawsze-czemu-mowimy-o-nim-teraz/>, (27.04.2019).
12 Michał Wróblewski, Wojciech Goszczyński, "Konflikt wokół monitoringu jakości powietrza w Polsce. Infrastruktury, standardy, dane," *Studia Socjologiczne*, Vol. 4, (2020), pp. 155–182. Michał Wróblewski, Wojciech Goszczyński, "Polish Smog Metrological Controversies," in: this volume.
13 Rogers, *Doing Digital Methods*, p. 21.
14 Rogers, *Doing Digital Methods*, p. 19.

order to be able to gather a spectrum of the most widely commented smog news in Poland, Google Alerts was used, a service that delivered a collection of news articles for the keywords: smog, air pollution, and smog Poland to an email inbox every day from the first of October 2019 to 01 May 2020.[15] The mapped area of smog controversy is therefore limited to the span of seven months in late 2019 and early 2020.

A browser plug-in, CrowdTangle, which collects information on the number of shares of a given article on Facebook, Instagram, and Twitter, was used to determine the intensity and breadth of a given controversy. The focus was primarily on Facebook shares due to the distribution of specific articles in Facebook groups rather than by individual users, which also allowed for political identification of specific groups. The data collected by CrowdTangle only relates to public Facebook groups, as this is currently the only data that can be collected from Facebook.[16] The mapping of controversies uses an empirical approach that does not separate epistemological from political controversies.[17]

From the data collected, a database was created in csv format, taking into account the source of the information provided, the five Facebook groups with the most Facebook reach, the number of Facebook reactions, the nature of the Facebook group in question, and the issue being discussed. This database was the basis for the controversy visualization that was created in RAWGraph using the Alluvial diagram

How to Detect Hotspots?

A method based on the study of controversies using digital methods, issue mapping, was used to locate the hotspots of debate.[18] Issue mapping, is a certain shift from controversy mapping,[19] which focuses on controversies understood as heated disputes around which there is shared uncertainty. Issue mapping, on the other hand, grows out of the assumption that it is rather the specific issues within a controversy that organize social life by focusing communities around

---

15 Piotr Idzik, "Analiza Big Data. Badanie niereaktywne w erze internetu 2.0," in: *Zwrot cyfrowy w humanistyce. Internet/Nowe Media/Kultura 2.0*, ed. Andrzej Radomski, Radosław Bomba, (Lublin: E-naukowiec, 2013), pp. 153–169.
16 Rogers, *Doing Digital Methods*, p. 19.
17 Marres, "Why map issues?," pp. 655–686.
18 Marres, "Why map issues?," pp. 655–686.
19 Venturini, "Diving in Magma," pp. 258–273; Venturini, "Building on faults," pp. 796–812.

them.[20] Moreover, issue mapping is a method adapted to the digital environment. As Noortje Marres points out, there is no controversy without media,[21] arguing that disputes in the digital environment should be analyzed by asking different questions than when analyzing controversies.[22] Issue mapping uses digital tools to explore controversies in online spaces. It is a method based on the belief that exploring problematic issues through a set of big data, allows for a better understanding of the dynamics of the "controversy" itself, but also of the technological infrastructures that are inherent in online debates.[23] Like other methods using digital tools, issue mapping also uses graphical visualizations to depict large data sets, as visualizations often allow for a narrative reading of the analyzed data.[24]

Controversy mapping is already a fairly well recognized practice within STS research. Researchers in this area have addressed the mapping of scientific controversies using a variety of research approaches e.g. the study of scientific controversies,[25] technological controversies[26] or the ethnography of the laboratory.[27] However, there is still relatively little work on Controversy Mapping using media messages, and even less that uses social media to do so. Controversy Mapping using social media is an area of research that has received the most attention

---

20 Noorjte Marres, Richard Rogers, "Recipe for Tracing Issues and Their Publics on the Web," in: *Making Things Public: Atmospheres of Democracy*, ed. Bruno Latour, Peter Weibel, (Cambridge: MIT Press, 2005), p. 927.
21 Noorje Marres, "No issues without media, The changing politics of public controversy in digital societies," in: *Media: A Transdisciplinary Inquiry*, ed. Jeremy Swartz, Janet Wasko, (Chicago: Intellect Books, 2021), pp. 1–25.
22 Marres, Rogers, "Recipe for Tracing," p. 5.
23 Marres, "Why map issues?," p. 663.
24 Tomasso Venturini, Liliana Bounegru, Mathieu Jacomy, Jonathan Gray, "How to Tell Stories with Networks. Exploring the Narrative Affordances of Graphs with the Iliad," in: *The Datafied Society Studying Culture through Data,* ed. Mirko T. Schafer, Karin van Es (Amsterdam: Amsterdam University Press, 2017), pp. 155–170.
25 Harry Collins, "The seven sexes: A study in the sociology of a phenomenon, or the replication of experiments in physics," *Sociology* Vol. 9, (1975) pp. 205–224; Harry Collins, "The Sociology of Scientific Knowledge: Studies of Contemporary Science," *Annual Review of Sociology,* Vol. 9, No.1 (2003), pp. 265–285.
26 Michael Bloor, Helen Sampson, Susan Baker, Katrin Dahlgren, "The Instrumental Use of Technical Doubts: Technological Controversies. Investment decisions and air pollution controls in the global shipping industry," *Science and Public Policy* Vol. 41, (2014), pp. 234–244.
27 Bruno Latour, Steve Woolgar, *Laboratory Life: The Constriction of Scientific Facts,* (Princeton: Princeton University Press, 1886).

to date from researchers associated with the University of Amsterdam.[28] Social media mapping makes it possible to show connections not only between actors themselves, but also between actors and specific content. Using Facebook or Twitter to depict the dynamics of a media controversy, allows for an accurate link between specific actors and their political background.[29] Some researchers assume that the media have little influence on the trajectory of controversies because they rarely address or transform strictly scientific claims. Thus, they sometimes influence the shape of the debate, but they rarely shape the knowledge that we consider valid. According to Noortje Marres, however, we cannot separate the mutual influence of the technological infrastructures used in litigation and the dynamics of controversy.[30] Methodological concerns about the use of social media in studying the dynamics of public debates, include the question of what are we mapping when we map a controversy using social media? Is it how they shape the debate, or how the media shape the debate? Or is it how the controversy shapes the media? Noortje Marees and David Moats answer: each of the above.[31] For we can both track how social media influences the content of the controversy, and how the controversy influences the role of social media.[32]

Smog Wars[33]

Issue mapping should begin by identifying the stabilized claims in its field. Within the smog controversy, one of the claims that appears most often in scientific, but also NGO or mainstream media messages is the belief that smog is harmful. The majority of smog-related information messages address the negative effects of air

---

28 Richard Rogers, Natalia Sánchez-Querubín, Aleksandra Kil, *Issue Mapping for an Ageing Europe*, (Amsterdam: Amsterdam University Press, 2015); Noortje Marres, David Moats, "Mapping Controversies with Social Media: The Case for Symmetry," *Social Media + Society*, (2015), pp. 1–17; Sabine Niederer, *Networked Content Analysis. The case of Climate Change*, (Amsterdam: Amsterdam University Press, 2016); Rogers, Marres, "Landscaping Climate Change."
29 Marres, "Why map issues?," p. 663.
30 Marres, "Why map issues?," p. 675.
31 Marres, Moats, "Mapping Controversies with Social Media," pp. 1–17.
32 Marres, Moats, "Mapping Controversies with Social Media," pp. 1–17.
33 The title of this subsection is taken from an article by Iwona Bojadżijewa, who refers to one of Krakow's anti-smog campaigns, see: Iwona Bojadżijewa, "Ekologia polityczna powietrza: o uwidzialnianiu miejskiego ryzyka," *Kultura i Społeczeństwo* Vol. 60, (2016), pp. 35–55.

pollution on our health as well as mortality rates. Scientific messages,[34] as well as reports from state institutions[35] and international institutions[36] focus mainly on this problem. We talk about smog when airborne substances harm humans, animals and plants. For this reason, we measure levels of PM10 and PM2.5, as well as benzopyrene, because they are the most harmful to health, despite the fact that there are more than 2,000 chemicals floating in the air.[37] The belief that smog is harmful has also been the basis for lawsuits by city governments and the state treasury. The municipality of Strzelce Opolskie was sued by inmates of Penitentiary No. 1 for insufficient actions to improve air quality in the city,[38] a lawsuit against the State Treasury was also brought by Krzysztof Waclawek, who suffered a carotid artery dissection and a mild stroke during a marathon run in Warsaw,[39] and a class action lawsuit by Warsaw residents against the State

---

34 Artur Badyda, Ewa Konduracka, Jakub Jędrak, "Wpływ zanieczyszczeń powietrza na umieralność," *SmogLab*, <https://smoglab.pl/wplyw-zanieczyszczen-powietrza-na-umieralnosc/>, (05.01.2020).
Joel Schwartz, Allan Marcus, "Mortality and Air Pollution in London: A Time Series Analysis," *American Journal of Epidemiology*, Vol. 131, (1990), pp. 185–194.
Bert Brunekreef, Stephen T. Holgate, "Air pollution and health," *Lancet*, Vol. 19, (2002), pp. 1233–1242.

35 NIK, "Dbaj o zdrowie – nie oddychaj. NIK o ochronie powietrza przed zanieczyszczeniami," (2018) <https://www.nik.gov.pl/aktualnosci/dbaj-o-zdrowie-nie-oddychaj.html>, (20.12.2019).
NIK, "Mordercze spaliny. NIK o ochronie powietrza przed zanieczyszczeniami cz. II. Emisja z sektora przemysłowego i transportu," (2018), <https://www.nik.gov.pl/aktualnosci/mordercze-spaliny.html> (20.12.2019); Główny Inspektorat Środowiska, "Badania zanieczyszczonego powietrza pyłem PM2,5 pod kątem monitorowania wskaźnika średniego narażenia," (2019), <http://powietrze.-gios.gov.pl/pjp/content/exposure_dust_pm>, (05.01.2020).

36 WHO, "Ambient air pollution;" WHO, "Household air pollution and health."

37 Maciej Bełcik, Katarzyna Piekarska, "Cytotoxicity properties of PM2.5 collected in Wrocław agglomeration," *E3S Web of Conferences*, Vol. 17, (2017), pp. 480–497.

38 Gazeta.pl, "Strzelce Opolskie. Więźniowie Zakładu Karnego pozwali gminę za smog nad spacerniakiem,"(2018),<https://wiadomosci.gazeta.pl/wiadomosci/7,114883,25509484,strzelce-opolskie-wiezniowie-zakladu-karnego-pozwali-gmine.html>, (17.03.2020).

39 Jakub Chełmiński, "Dostał udaru podczas maratonu. Twierdzi, że przez smog. Teraz organizuje protest,"(2019),*Gazeta Wyborcza*,<https://warszawa.wyborcza.pl/warszawa/7,54420,24483928,dostal-udaru-podczas-maratonu-twierdzi-ze-przez-smog-teraz.html>, (05.01.2020).

Treasury is also pending.[40] The first judgment with a verdict against the city government was issued in February 2020.[41]

Over the span of seven months, ten of the most widely commented smog-related disputes in Poland were located in the smog controversy area. These controversies can be divided into ontological – concerning the status of smog in our collective, and epistemological – concerning the negotiation of proposed ways to improve air quality. The ontological controversy was the most widely commented and received the most shares on social media, especially on politically declared and conspiratorial groups. Empirical controversies, on the other hand, received much less publicity, and were shared primarily by popular science groups, grassroots NGO initiatives, or official government websites.

The debate with the greatest reach on social media concerned an alternative smog map that was supposed to be evidence of good air quality in Poland, compared to Western countries, which, according to the visual representation, do much worse with urban air pollution. The map was first published by the *Twoja Pogoda* portal in an article with the telling title: "Poland most polluted in Europe? New smog map reveals it's complete nonsense,"[42] and then copied into news stories on other, slightly more popular news portals. The themes of the other published articles were in a similar mood, demystifying denunciations of poor air quality in Poland.[43] What was also surprising was the popularity of this information, on Twitter, where this information was spread by Polish politicians,

---

40 Marcin Śmigiel, "223 osoby pozwały państwo za smog. Będą odszkodowania za leczenie czy zakup maseczki antysmogowej?," (2018), <https://warszawa.wyborcza.pl/warszawa/7,54420,25240192,223-osoby-pozwaly-panstwo-za-smog-beda-odszkodowania-za-koszty.html>, (05.01.2020).

41 Marcin Śmigiel, "Pierwszy taki wyrok: miasto przegrało w sądzie bo nieskutecznie walczy ze smogiem," Gazeta Wyborcza (2020), <https://warszawa.wyborcza.pl/warszawa/7,54420,25705731,pierwszy-taki-wyrok-miasto-przegralo-w-sadzie-bo-nieskutecznie.html>, (18.02.2020).

42 Twoja Pogoda, "Polska najbardziej skażona w Europie? Nowa mapa smogowa ujawnia, że to kompletna bzdura," (2019), <https://www.twojapogoda.pl/wiadomosc/2019-12-13/polska-najbardziej-skazona-w-europie-nowa-mapa-smogowa-ujawnia-ze-to-kompletna-bzdura/>, (17.12.2019).

43 TVP Info, "Najnowsza mapa smogowa jest dla nas łaskawa," (2019), <https://www.tvp.info/45807151/najnowsza-mapa-smogowa-jest-dla-nas-laskawa-w-czolowce-trucicieliikony-ekologii>, (17.12.2019); Portal samorządowy, "Mapa smogu. KE przedstawia zaskakujące dane na temat zanieczyszczenia powietrza," (2019), <https://www.portalsamorzadowy.pl/ochrona-srodowiska/mapa-smogu-ke-przedstawila-zaskakujace-dane-na-temat-zanieczyszczenia-powietrza,139869.html>, (17.12.2019).

including the President of Poland Andrzej Duda,[44] but also the MP of the Civic Coalition Paulina Henning-Kloski, the member of the Committee on Industry, Research and Energy of the European Parliament Jacek Saryusz-Wolski, or the chairman of the Committee on National Economy and Innovation, PiS Senator Marek Pęka. An article on the portal *Twoja Pogoda* was described by Jakub Jędrak, member of the Polish Smog Alarm, as manipulation.[45]

The controversy about air pollution levels in Poland was also popular on social media. The debate began with an article published by Greenpeace's Unearthead website, which cited WHO's summary of urban air pollution.[46] However, the information provided on this topic in the mainstream media in Poland was variously oriented.[47] Finally, the controversy about the level of exhaust air pollution in Polish cities erupted during the first lockdown related to the coronavirus outbreak, in late March and early April 2020. At that time, many news portals published articles on the minimal impact of car exhaust on air pollution levels in Poland.[48]

The much more numerous epistemological controversies did not garner such wide coverage on social media. However, they were characterized by a longer duration and focused a larger number of actors who took part in negotiations about potential solutions to the smog problem in Poland. They were also controversies that often led to significant transformations of knowledge about air

---

44 President Andrzej Duda Twitter Account, <https://twitter.com/MarcinPalade/status/1205628417133162496>, (17.12.2019).

45 Jakub Jędrak, "Nowa 'mapa smogowa' ujawnia, jak łatwo sprzedać ludziom bzdury," (2019), <https://smoglab.pl/nowa-mapa-smogowa-ujawnia-jak-latwo-sprzedac-ludziom-bzdury/>, (17.12.2019).

46 Lauri Myllyvirta, Emma Howard, "Five things we learned from the world's biggest air pollution database," (2018), <https://unearthed.greenpeace.org/2018/05/02/air-pollution-cities-worst-global-data-world-health-organisation/>, (06.11.2019).

47 TVP Info, "Polskie miasta w czołówce najbardziej zanieczyszczonych w Europie," (2018), <https://www.tvp.info/39611492/polskie-miasta-w-czolowce-najbardziej-zanieczysz- czonych-w-europie>, (06.11.2019); TVP Info, "Najnowsza mapa smogowa jest dla nas łaskawa;" Artur Stando, "Najnowszy raport 50 najbardziej zanieczyszczonych miast. Polska dominuje," (2018), <https://tech.wp.pl/najnowszy-ranking-50-najbardziej-zanieczyszczonych-miast-pol- ska-dominuje-6249269556491905a>, (06.11.2019).

48 Focus, "Kwarantanna i jej zbawienny wpływ na jakość powietrza. Tylko w Polsce wciąż smog," (2020),<https://www.focus.pl/artykul/kwarantanna-i-jej-zbawienny-wplyw-na-jakosc-powietrza-eea-pokazuje-dane>, (27.03.2020); Onet motoryzacja, "Nie ma aut na ulicach, jest smog. Jak to możliwe?" (2020), <https://www.auto-swiat.pl/porady/nie-ma-aut-na-ulicach-jest-smog-jak-to-mozliwe/924h22p>, (23.03.2020).

pollution in Poland. One of the longest-running epistemological controversies included the debate between the Polish Smog Alarm (PAS) and the Ministry of Climate and the Ministry of Environment. The reduction by the Ministry of Climate in October 2019 of the information and alert levels for particulate air pollution is the result of negotiations led by PAS. The permissible level of PM10 pollution, as an alert level, was halved from 300 micrograms per cubic meter to 150 micrograms per cubic meter, while the applicable pollution notification level was reduced from 200 to 100 micrograms per cubic meter. PAS already in 2016 questioned the values of air pollution accepted in Poland as acceptable, comparing them with the limit values in other European countries. The authors of the article on air pollution standards drew attention to the values set by the World Health Organization, it considers as acceptable: 25 μg/m3 for PM 2.5 and 50 μg/m3 for PM 10. At that time, air pollution levels of PM2.5 and PM10 were 6–8 times higher for PM2.5 and 8–10 times higher for PM10.

A discussion started by PAS on the performance of one of the government's first smog-fighting programs, Clean Air, also resulted in a transformation of the program. Several months into the Clean Air Program, PAS produced a report summarizing its initial form (PAS, 2020). It highlighted the main problems with the program, due to which the rate of replacement of outdated stoves was extremely slow – initial estimates assumed replacement of all non-compliant heating boilers in Polish cities within a several decades, and in some locations, e.g. Łódź, even for one hundred and twenty years.[49] The authors of the PAS report also drew attention to the financial regulations which, in essence, have so far excluded the poorest persons from the programme. The problem with the Clean Air Programme was publicised by mainstream media[50] due to the reaction of the European Commission, which threatened Poland with discontinuation of financing the programme due to its inefficiency.

Anti-smog laws were also controversial, widely consulted with residents, local authorities, doctors, boiler manufacturers, representatives of the national chamber of chimney sweeps, and NGOs. The anti-smog laws are the beginning of an extensive controversy concerning both the technology allowed for heating

---

49 PAS, "Likwidacja 'kopciuchów' stoi w miejscu."
50 Danuta Wantuch, "Miliardy euro nie dotrą do nas. Polska dalej będzie zasnuta smogiem," Gazeta Wyborcza (2019), <https://wyborcza.pl/7,155287,24949458,miliardy-euro-nie-do- tra-do-polski.html>, (09.03.2020); Spider's Web, "Bez przełomu w sprawie Czystego Powietrza. Co dalej z finansowaniem programu?," (2020), <https://spidersweb.pl/bizblog/czyste-powietrze-komisja-europejska/>, (08.03.2020).

homes, the affluence of residents in particular provinces, the price of Polish coal, or the ways to control the level of low emissions. The first law voted in 2017 for the Małopolskie Voivodeship,[51] initially concerned only the obligation to install stoves and boilers meeting EU eco-design standards, as well as the replacement of coal-fired boilers not meeting any standards. Resolutions have now been adopted for ten, out of sixteen voivodeships in Poland.

In turn, the controversy around climate taxes started with the city of Zakopane, which was sued by a tourist in 2018 for charging a climate fee despite very poor air quality during the winter season.[52] The case was decided against the city, even after the city authorities appealed to the Court of Cassation. Zakopane as of 2018 can no longer charge a climate fee. In 2019, Beniamin Łuczyński – member of the association *Miasto jest nasze,* and Patryk Białas – councillor of Katowice, member of the PO, activist in the association *BoMiasto*, sued Sandomierz, Toruń and Szczyrk for collecting local fees.[53] The activists, as well as representatives of the Client Earth foundation, referred to the Administrative Court ruling in the Zakopane case. Kraków and Sandomierz also opted out of the climate fee due to poor air quality.[54]

Issue Mapping

The visual representation of the mapped debate around smog in Poland over the seven months from October 2019 to May 2020, shown below, captures the main areas of smog controversy and their reach, as well as the Facebook groups that shared particular links. The Alluvial diagram shows the issue taken up by websites, and the Facebook groups that published specific links. The wider the link between the link and the Facebook group, the more shares and interactions the link received. The wider the link between the problem and the link, the more

---

51 PAS, "Pierwsza w Polsce uchwała antysmogowa przyjęta w Małopolsce," (2017), <https://polskialarmsmogowy.pl/polski-alarm-smogowy/aktualnosci/szczegoly,pierwsza-w-polsce-uchwala-antysmogowa-przyjeta-w-malopolsce,355.html>, (16.03.2020).

52 TOK FM, "Turysta wygrał z Zakopanem proces o opłatę klimatyczną," (2018), <https://www.tokfm.pl/Tokfm/7,130517,23147500,turysta-wygral-z-zakopanem-proces-o-oplate-klimatyczna-przez.html>, (17.03.2020).

53 Onet.pl, "Turyści skarżą opłaty klimatyczne w kolejnych polskich miastach. Płace tak naprawdę za straszliwy smog," (2018), <https://podroze.onet.pl/polska/oplaty-klimatyczne-w-polskich-miastach-a-smog/27rn433>, (17.03.2020).

54 Radio Kielce, "Sandomierz zlikwidował opłatę klimatyczną," (2020), <http://www.ra-dio.kielce.pl/pl/post-98630>, (17.03.2020).

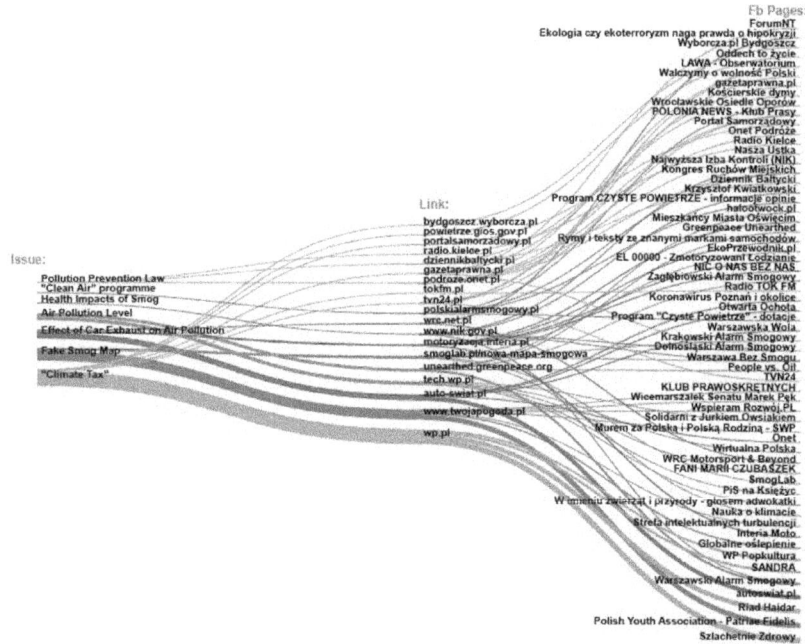

Figure 1. Alluvial chart: relationships between issues, links and groups on Facebook
Source: own elaboration.

links were related to the same problem. The colors correspond to the specific issues addressed in the online debates.

Most of the press articles published and links shared on Facebook were about the climate tax, the fake smog map, and the impact of car exhaust on air pollution. Little space was devoted to debates about air pollution levels in Poland, smog's impact on health, or government programs that were supposed to partially deal with the smog problem. It might seem that this effect is related to the relationship between news portals (links) and the Facebook groups that share them. However, this relationship only exists between science portals and scientific or non-governmental Facebook groups, because in both mainstream media and social media, debates about local smog issues were most often shared by groups associated with the clean air campaign (local smog alerts). For the remaining "issues," however, it was more likely issues that organized Facebook groups around

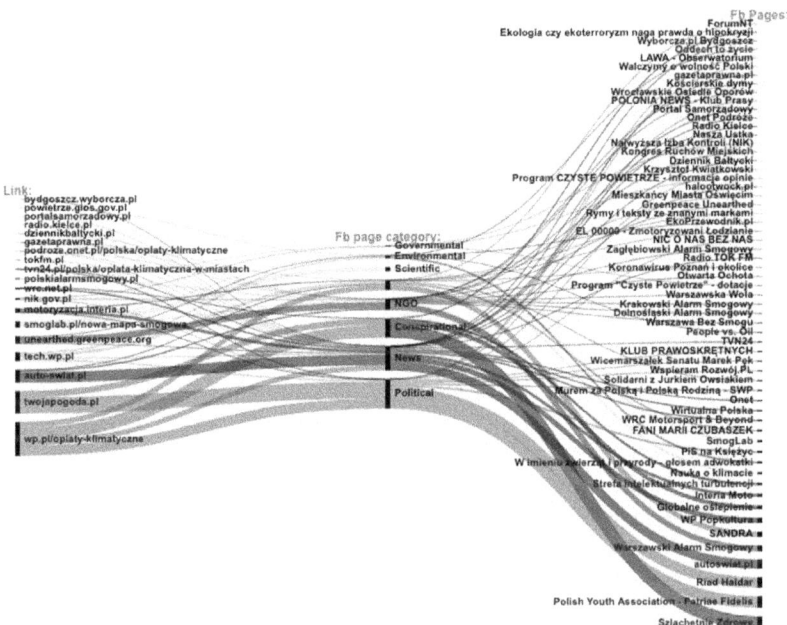

Figure 2. Alluvial chart: relations between links, Facebook groups and their category
Source: own elaboration.

them. Thus, it is issues that organize the realm of Facebook groups, rather than links that are then shared by groups.

The second graph shows the relationship between shared links, Facebook groups, and public Facebook page categories. Coding of Facebook groups was conducted based on descriptions of groups on Facebook, as well as content analysis of pages. The vast majority of groups are scientific groups (*Nauka dla klimatu, W imieniu zwierząt i przyrody*) and non-governmental groups (Greenpeace, local smog alerts), a large proportion of groups are also Facebook groups of large news portals (*TVN, WP, Onet*). A small number of groups are political groups (*KOD, PiS na Księżyc, Klub prawoskrętnych*). The biggest number of engagements belongs to political and conspiracy groups, despite the fact that conspiracy groups are only *Szlachetnie zdrowy* and *Globalne oślepienie*. The former is a group dedicated to alternative medicine, while the latter is a climate denialist site.

The links that were most widely discussed – *twojapogoda.pl* and *wp.pl* domains – were distributed in groups of a political and conspiratorial nature. These are the links that were most related to the most widely discussed issues – fake smog map and climate tax (both related to the debate about the future of coal-based economy in Poland). However, the domains that published articles on these topics are not political or conspiratorial sites. *Twoja Pogoda* is a weather forecasting website and *WP* is one of the larger news portals in Poland. Rather, the topics on which the articles were published were intercepted by Facebook groups, who then published the given links with their own commentary. The *Twoja Pogoda* portal published a fake smog map, and *WP* published texts on climate tax, which sparked discussions mainly in politically diverse groups. Despite the fact that most of the issues around the topic of smog in Poland were debates concerning potential solutions to the smog problem, the issues that gained the most engagement were those related to undermining the problem of air pollution in Poland and undermining economic solutions that would partly solve the problem of smog in Poland. Such messages were associated with specific groups that were politically or conspiracy oriented. Despite the clear involvement of scientific, governmental, and NGO groups in the discussion, the data visualization clearly indicates their low visibility in the public discourse. The highest engagements were gained by groups publishing political and conspiratorial content. In this case, seemingly asymmetrical actors, i.e. those with incommensurate relationships or stronger allies, often do not have such broad media reach. Often, actors who appear to be less influential produce a debate with far greater reach than, for example, governmental institutions.

Conclusion

The results of this analysis are, of course, fragmentary. One of the objections directed towards research using digital methods is the bias of the data on which they are based.[55] The reaction to these accusations has been the development of two approaches: precautionary and affirmative. The precautionary approach assumes that we should aim for the "cleanest" data possible on which to conduct controversy analyses. The affirmative approach relies on data that is always biased to some degree, using it to examine the relationship between the technological infrastructure we use and the dynamics of the controversy we receive.[56]

---

55 Marres, "Why map issues?."
56 Marres, "Why map issues?," pp. 665–667.

Similarly, and in this case, we should first ask how the data generated by Google Alert is filtered by this overlay? And what vision of the debate around smog does this software produce? How does analysis through the Facebook platform affect the dynamics of this controversy?

Google Alert is based on a keyword search that we enter when creating an alert. As a result, the results we get are narrowed down to those that have one or all of the keywords we selected. In addition, Google Alerts do not always work completely smoothly, often overlooking some of the results. The number of articles, and the links between them, made it easy in most cases to find those to which others referred. However, it only reflects newspaper articles that were published on the topic, not providing results containing keywords on other platforms. Hence this is the dynamics of the debate provided by public media. Tracking the fate of links from public media and their transformation on Facebook, however, allowed us to reconstruct the dynamics of mainstream media circulation on Facebook. Although the link base was limited to that of Google Alert, their fluctuation on Facebook looks different. In the "Smog Wars" subsection, the dynamics of the controversy that was briefly described look quite different than the analysis based on the reach garnered by news articles. The most widespread controversy in both cases remains the issue of the fake smog map, but press coverage would suggest that the biggest part of the public debate over smog is the negotiations between PAS and the Polish government. However, analysis of the reach of individual articles shows that the topics that gained the most shares on Facebook were those related to questioning the existence of smog in Poland and discussions about the Climate Tax.

The designed study could have been conducted using other approaches – tracking URLs links, hashtags, keywords, or comparing debate across platforms. This study covers only a small part of the smog discourse in Poland in late 2019 and early 2020. It would be interesting to see the results of a comparison of the arguably diverse dynamics of this controversy depending on the platform and method adopted to study media messages about smog in Poland. Such a study would allow for a better understanding of the influence of technological infrastructures on the shape of the public debate on smog in Poland.

Bibliography

Badyda, Artur, Ewa Konduracka, Jakub Jędrak. "Wpływ zanieczyszczeń powietrza na umieralność." SmogLab. 2016. https://smoglab.pl/wplyw-zanieczyszczen-powietrza-na-umieralnosc/ (20 Apr. 2022).

Beck, Gerald and Cordula Kropp. "Infrastructures of risk: A mapping approach toward controversies on risks." *Journal of Risk Research* Vol. 14, 1/2011, pp. 1–16.

Bełcik, Maciej and Katarzyna Piekarska. "Cytotoxicity properties of PM2.5 collected in Wrocław agglomeration." *E3S Web of Conferences* 17/2017, pp. 480–497.

Bińczyk, Ewa. "Doktor Golem? Badania kontrowersji w nurcie studiów nad nauką i technologią wobec statusu nauk medycznych, wybranych jednostek chorobowych i zmieniającej się roli eksperta." *Przegląd Filozoficzno-Literacki*, Vol. 1, 46/2017, pp. 285–306.

Bloor, Michael, Helen Sampson, Susan Baker and Katrin Dahlgren. "The Instrumental Use of Technical Doubts: Technological Controversies. Investment decisions and air pollution controls in the global shipping industry." *Science and Public Policy* 41/2014, pp. 234–244.

Bojadżijewa, Iwona. "Ekologia polityczna powietrza: o uwidzialnianiu miejskiego ryzyka." *Kultura i Społeczeństwo* 60(2)/2016, pp. 35–53.

Brunekreef, Bert and Stephen T. Holgate. "Air pollution and health." *Lancet* 19/2002, pp. 1233–1242.

Chełmiński, Jakub. "Dostał udaru podczas maratonu. Twierdzi, że przez smog. Teraz organizuje protest." GazetaWyborcza. 2019. https://warszawa.wybor cza.pl/warszawa/7,54420,24483928,dostal-udaru-podczas-maratonu-twier dzi-ze-przez-smog-teraz.html. (5 Jan. 2020).

Collins, Harry and Trevor Pinch. *The Golem: What Everyone Should Know about Science*. Cambridge: Cambridge University Press, 1993.

Collins, Harry. "The seven sexes: A study in the sociology of a phenomenon, or the replication of experiments in physics." *Sociology* 9/1975, pp. 205–224.

Collins, Harry. "The Sociology of Scientific Knowledge: Studies of Contemporary Science." *Annual Review of Sociology* Vol. 9, 1/2003, pp. 265–285.

Focus. "Kwarantanna i jej zbawienny wpływ na jakość powietrza. Tylko w Polsce wciąż smog." 2020. https://www.focus.pl/artykul/kwarantanna-i-jej-zbawie nny-wplyw-na-jakosc-powietrza-eea-pokazuje-dane (27 Mar. 2020).

Gazeta.pl. "Strzelce Opolskie. Więźniowie Zakładu Karnego pozwali gminę za smog nad spacerniakiem." 2018. https://wiadomosci.gazeta.pl/wiadomosci/ 7,114883,25509484,strzelce-opolskie-wiezniowie-zakladu-karnego-pozwali-gmine.html (17 Mar. 2020).

Germani, Federico and Nikola Biller-Andorno. "The anti-vaccination infodemic on social media: A behavioral analysis." *PLoS ONE* Vol. 16(3)/2021.

Główny Inspektorat Środowiska. *"Badania zanieczyszczonego powietrza pyłem PM2,5 pod kątem monitorowania wskaźnika średniego narażenia."* 2019. http://powietrze.-gios.gov.pl/pjp/content/exposure_dust_pm (05 Jan. 2020).

Idzik, Piotr. "Analiza Big Data. Badanie niereaktywne w erze internetu 2.0." In: *Zwrot cyfrowy w humanistyce. Internet/Nowe Media/ Kultura 2.0*, ed. Andrzej Radomski and Radosław Bomba. Lublin: E-naukowiec, 2013, pp. 153–169.

Jędrak, Jakub. "Nowa 'mapa smogowa' ujawnia, jak łatwo sprzedać ludziom bzdury." 2019. https://smoglab.pl/nowa-mapa-smogowa-ujawnia-jak-latwo-sprzedac-lu-dziom-bzdury/ (17 Dec. 2019).

Kuhn, Tomas. *The Structure of Scientific Revolution*. Chicago: University of Chicago Press, 1962.

Latour, Bruno and Steve Woolgar. *Laboratory Life: The Constriction of Scientific Facts*. Princeton: Princeton University Press, 1986.

Mager, Astid and Jenny Eklöf. "Technoscientific Promotion and Biofuel Policy: How the Press and Search Engines Stage the Biofuel Controversy." *Media Culture & Society* Vol. 35, 4/2013, pp. 454–471.

Marres, Noortje. "Why map issues? On Controversy Analysis as a Digital Method." *Science, Technology, & Human Values* Vol. 40, 5/2015, pp. 655–686.

Marres, Noorje. "No issues without media, The changing politics of public controversy in digital societies." In: *Media: A Transdisciplinary Inquiry*, ed. Jeremy Swartz and Janet Wasko. Chicago: Intellect Books, 2021, pp. 1–25.

Marres, Noortje and David Moats. "Mapping Controversies with Social Media: The Case for Symmetry." *Social Media & Society* 1/2015, pp. 1–17.

Marres, Noorjte and Richard Rogers. "Recipe for Tracing Issues and Their Publics on the Web." In: *Making Things Public: Atmospheres of Democracy*, ed. Bruno Latour and Peter Weibel. Cambridge: MIT Press, 2005, p. 927.

Mukerji, Chandra. "Controversy studies." In: *The Blackwell Encyclopedia of Sociology*, ed. Georg Ritzer. Oxford: John Wiley & Sons, 2007, pp. 784–788.

Myllyvirta, Lauri, and Emma Howard. "Five things we learned from the world's biggest air pollution database." 2018. https://unearthed.greenpeace.org/2018/05/02/air-pollution-cities-worst-global-data-world-health-organisation/ (06 Nov 2019).

Niederer, Sabine. *Networked Content Analysis. The case of Climate Change*. Amsterdam: Amsterdam University Press, 2016.

Niederer, Sabine. "Global warming is not a crisis!: Studying climate change skepticism on the Web." *European Journal of Media Studies* Vol. 2, 1/2013, pp. 83–112.

NIK. "Dbaj o zdrowie – nie oddychaj. NIK o ochronie powietrza przed zanieczyszczeniami." 2018. https://www.nik.gov.pl/aktualnosci/dbaj-o-zdrowie-nie-oddychaj.html/ (20 Dec. 019).

NIK. "Mordercze spaliny. NIK o ochronie powietrza przed zanieczyszczeniami cz. II. Emisja z sektora przemysłowego i transportu." 2018. https://www.nik.gov.pl/aktualnosci/mordercze-spaliny.html/ (20 Dec. 2019).

Onet motoryzacja. "Nie ma aut na ulicach, jest smog. Jak to możliwe?" 2020. https://www.auto-swiat.pl/porady/nie-ma-aut-na-ulicach-jest-smog-jak-to-mozliwe/924h22p/ (23 Mar. 2020).

Onet.pl. "Turyści skarżą opłaty klimatyczne w kolejnych polskich miastach. Płace tak naprawdę za straszliwy smog." 2018. https://podroze.onet.pl/polska/oplaty-klima-tyczne-w-polskich-miastach-a-smog/27rn433/ (17 Mar. 2020).

Oreskes, Naomi and Erik M. Conway. *Merchants of Doubts. How a Handful of Scientist Obscured the Truth on Issues from Tabaco Smoke to Global Warming.* New York: Bloomsbury Press, 2010.

PAS. "Pierwsza w Polsce uchwała antysmogowa przyjęta w Małopolsce." 2017. https://polskialarmsmogowy.pl/polski-alarm-smogowy/aktualnosci/szczegoly,pierwsza-w-polsce-uchwala-antysmogowa-przyjeta-w-malopolsce,355.html/ (16 Mar. 2020).

PAS. "Poziomy informowania i alarmowe." https://ww-w.polskialarmsmogowy.pl/polski-alarm-smogowy/smog/szczegoly,poziomy-informowania-i-alarmowe,19.html/ (07 Feb. 2020).

PAS. "Likwidacja 'kopciuchów' stoi w miejscu – PAS podsumowuje program wymiany kotłów." 2020. https://polskialarmsmogowy.pl/polski-alarm smogowy/aktualnosci/szczegoly,likwidacja-kopciuchow-stoi-w-miejscu---pas-podsumowuje-programy-wymiany-kotlow,1403.html/ (4 Mar. 2020).

Portal samorządowy. "Mapa smogu. KE przedstawia zaskakujące dane na temat zanieczyszczenia powietrza." 2019. https://www.portalsamorzadowy.pl/ochrona-srodowiska/mapa-smogu-ke-przedstawila-zaskakujace-dane-na-temat-zanieczyszczenia-powietrza,139869.html/ (17 Dec. 2019).

President Andrzej Duda Twitter Account. https://twitter.com/MarcinPalade/status/1205628417133162496/ (17 Dec. 2019).

Radio Kielce. "Sandomierz zlikwidował opłatę klimatyczną." 2020. http://www.ra-dio.kielce.pl/pl/post-98630/ (17 Mar. 2020).

Rogers, Richard and Noortje Marres. "Landscaping climate change: a mapping technique for understanding science and technology debates on the World Wide Web." *Public Understanding of Science.* 9/2000, pp. 1–23.

Rogers, Richard, Natalia Sánchez-Querubín and Aleksandra Kil. *Issue Mapping for an Ageing Europe.* Amsterdam: Amsterdam University Press, 2015.

Rogers, Richard. *Doing Digital Methods*. Los Angeles: Sage, 2019.

Schwartz, Joel and Allan Marcus. "Mortality and Air Pollution in London: A Time Series Analysis." *American Journal of Epidemiology*. 131/1990, pp. 185–194.

Smoglab. "Smog był zawsze, czemu mówimy o nim teraz?" https://smoglab.pl/smog-byl-zawsze-czemu-mowimy-o-nim-teraz/ (27 Apr. 2019).

Spider's Web. "Bez przełomu w sprawie Czystego Powietrza. Co dalej z finansowaniem programu?" 2020. https://spidersweb.pl/bizblog/czyste-powietrze-komisja-europejska/ (8 Mar. 2020).

Stando, Artur. "Najnowszy raport 50 najbardziej zanieczyszczonych miast. Polska dominuje." 2018. https://tech.wp.pl/najnowszy-ranking-50-najbardziej-zanieczyszczonych-miast-polska-dominuje-6249269556491905a/ (06 Nov. 2019).

Śmigiel, Marcin. "223 osoby pozwały państwo za smog. Będą odszkodowania za leczenie czy zakup maseczki antysmogowej?" 2018. https://warszawa.wyborcza.pl/warszawa/7,54420,25240192,223-osoby-pozwaly-panstwo-za-smog-beda-odszkodowania-za-koszty.html/ (05 Jan. 2020).

Śmigiel, Marcin. "Pierwszy taki wyrok: miasto przegrało w sądzie bo nieskutecznie walczy ze smogiem." Gazeta Wyborcza. 2020. https://warszawa.wyborcza.pl/warszawa/7,54420,25705731,pierwszy-taki-wyrok-miasto-przegralo-w-sadzie-bo-nieskutecznie.html/ (18 Feb. 2020).

TOK FM. "Turysta wygrał z Zakopanem proces o opłatę klimatyczną." 2018. https://www.tokfm.pl/Tokfm/7,130517,23147500,turysta-wygral-z-zakopanem-proces-o-oplate-klimatyczna-przez.html/ (17 Mar. 2020).

TVP Info. "Polskie miasta w czołówce najbardziej zanieczyszczonych w Europie." 2018. https://www.tvp.info/39611492/polskie-miasta-w-czolowce-najbardziej-zanieczyszczonych-w-europie/ (06 Nov. 2019).

TVP Info. "Najnowsza mapa smogowa jest dla nas łaskawa." 2019. https://www.tvp.info/45807151/najnowsza-mapa-smogowa-jest-dla-nas-laskawa-w-czolowce-trucicieliikony-ekologii/ (17 Dec. 2019).

Twoja Pogoda. "Polska najbardziej skażona w Europie? Nowa mapa smogowa ujawnia, że to kompletna bzdura." 2019. https://www.twojapogoda.pl/wiadomosc/2019-12-13/polska-najbardziej-skazona-w-europie-nowa-mapa-smogowa-ujawnia-ze-to-kompletna-bzdura/ (17 Dec. 2019).

Venturini, Tomasso. "Diving in Magma: How to Explore Controversies with Actor – Network Theory." *Public Understanding of Science*, 19/2010, pp. 258–273.

Venturini, Tomasso. "Building on faults: how to represent controversies with digital methods." *Public Understanding of Science* 21/2012, pp. 796–812.

Venturini, Tomasso, Liliana Bounegru, Mathieu Jacomy and Jonathan Gray. "How to Tell Stories with Networks. Exploring the Narrative Affordances of Graphs with the Iliad." In: *The Datafied Society Studying Culture through Data*, ed. Mirko T. Schafer, Karin van Es. Amsterdam: Amsterdam University Press, 2017, pp. 155–170.

Venturini, Tomasso and Daniele Guido. "Once Upon a Text: an ANT Tale in Text Analysis." *Sociologica*. 3/2012, pp. 1–16.

Wantuch, Danuta. "Miliardy euro nie dotrą do nas. Polska dalej będzie zasnuta smogiem." Gazeta Wyborcza. 2019. https://wyborcza.pl/7,155287,24949458,miliardy-euro-nie-dotra-do-polski.html/ (09 Mar. 2020).

WHO. "Ambient air pollution." https://www.who.int/airpollution/data/cities/en/ (05 Jan. 2020).

WHO. "Household air pollution and health." https://www.who.int/news-room/fact-sheets/detail/household-air-pollution-and-health/ (01 May. 2020).

Wróblewski Michał, Goszczyński Wojciech. "Konflikt wokół monitoringu jakości powietrza w Polsce. Infrastruktury, standardy, dane." *Studia Socjologiczne* 4/2020, pp. 155–182.

Comparative Studies on Education, Culture and Technology
Vergleichende Studien zur Bildung, Kultur und Technik

Edited by / Herausgegeben von
Tomasz Stępień

Vol. / Bd. 1   Tomasz Stępień / Annette Deschner / Mojca Kompara / Adriana Merta-Staszczak: Spatialisation of Education. Migrating Languages – Cultural Encounters – Technological Turn. 2013.

Vol. / Bd. 2   Anton Hilckman: Gesammelte Werke. Schriften zur philosophischen Pädagogik Teil 1. Bildung – Begeisterung – Freiheit. Bearbeitet, kommentiert und herausgegeben von Tomasz Stępień. 2014.

Vol. / Bd. 3   Anton Hilckman: Gesammelte Werke. Schriften zur philosophischen Pädagogik Teil 2. Christliche Philosophie. Bearbeitet, kommentiert und herausgegeben von Tomasz Stępień. 2014.

Vol. / Bd. 4   Ewa Bińczyk / Tomasz Stępień: Modeling Technoscience and Nanotechnology Assessment. Perspectives and Dilemmas. 2014.

Vol. / Bd. 5   Tomasz Stępień: Heuristics of Technosciences. Philosophical Framing in the Case of Nanotechnology. 2016.

Vol. / Bd. 6   Mojca Kompara / Tomasz Stępień: Spatialisation of Higher Education: Poland and Slovenia. 2017.

Vol. / Bd. 7   Hilckman Anton: Gesammelte Werke. Schriften zur politischen Pädagogik. Teil 1: Politische Theorie und Föderalismus Bearbeitet, kommentiert und herausgegeben von Tomasz Stępień. 2019.

Studies on Culture, Technology and Education

Edited by Krzysztof Abriszewski

Vol. / Bd. 8   Krzysztof Abriszewski / Aleksandra Derra / Andrzej W. Nowak (eds.): Polish Science and Technology Studies in the New Millennium. 2022.

www.peterlang.com

www.ingramcontent.com/pod-product-compliance
Ingram Content Group UK Ltd.
Pitfield, Milton Keynes, MK11 3LW, UK
UKHW021829210426
5322IPUK00004B/90